Confederate Saboteurs

ED RACHAL FOUNDATION NAUTICAL ARCHAEOLOGY SERIES
in association with the Institute of Nautical Archaeology

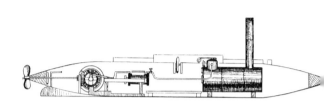

Mark K. Ragan

Confederate Saboteurs

Building the *Hunley*
and Other Secret Weapons
of the Civil War

Texas A&M University Press
College Station

Copyright © 2015 by Mark K. Ragan
All rights reserved
First edition

This paper meets the requirements of
ANSI/NISO Z39.48-1992
(Permanence of Paper).
Binding materials have been chosen for durability.
Manufactured in the United States of America

Library of Congress Cataloging-in-Publication Data

Ragan, Mark K., author.
Confederate saboteurs: building the Hunley and other secret weapons of the Civil War / Mark K. Ragan. — First edition.
pages cm — (Ed Rachal Foundation nautical archaeology series)
Includes bibliographical references and index.
ISBN 978-1-62349-278-6 (cloth: alk. paper) — ISBN 978-1-62349-279-3 (ebook) 1. Confederate States of America. Singer Secret Service Corps—Biography. 2. Confederate States of America. Singer Secret Service Corps—History. 3. Singer, Edgar Collins, 1825–1919. 4. Confederate States of America. Army—Officers—Biography. 5. Mechanical engineers—Biography. 6. Secret service—Confederate States of America—History. 7. United States—History—Civil War, 1861–1865—Naval operations—Submarine. 8. H.L. Hunley (Submarine) I. Title. II. Series: Ed Rachal Foundation nautical archaeology series.
E467.R34 2015
973.7'86—dc23
2014046006

This book is dedicated to Clive Cussler and his NUMA organization. After searching for the final resting place of the *H. L. Hunley* for nearly fifteen years, their unwavering persistence and dedication to the quest was finally rewarded in early May of 1995.

Contents

A gallery of images follows page 98.

   Introduction · 1
1. Formation and Deployment · 8
2. First Torpedo Strike and Launching the *Hunley* · 38
3. Richmond Invests in the Corps · 61
4. Losing the *Hunley* · 99
5. Operations throughout the Confederacy · 134
6. Taking the Fight to the Enemy · 165
   Conclusion · 185
   Notes · 193
   Bibliography · 227
   Index · 241

Confederate Saboteurs

Introduction

OF THE VARIOUS TORPEDO CORPS and secret service companies organized and deployed by the Confederate government during the American Civil War, none was more energetic, ingenious, or successful than the group headed by Texan Edgar Collins Singer. With an assailable coastline of thirty-five hundred miles and numerous waterways leading deep into the South's heartland, the newly formed Confederacy had difficulty defending its self-proclaimed territory from the vastly superior Federal government forces. The largely agrarian South, which had few established industries, was forced to turn to unorthodox forms of warfare in the struggle for independence, one of which was the government financing of destructive underwater devices.

The Singer Secret Service Corps (also known as the Singer Submarine Corps and Singer's Torpedo Company) was founded by a small group of middle-aged Masons at Port Lavaca, Texas, in the early spring of 1863. The twenty-five to thirty initiated members of this secret organization were from diverse occupational backgrounds, ranging from steam engineers and attorneys to machinists, jewelers, tinkers, and gunsmiths. During the last two years of the war this group of engineers, agents, and operatives would develop and deploy inventions such as huge ironclad torpedo boats, submarine vessels, underwater weaponry, and explosive devices far superior to anything that had been seen up to the mid-nineteenth century.

The group's main government-financed activity was the manufacture and deployment of an underwater contact mine that had been perfected and patented by Edgar C. Singer of Shea's Battery of Texas Light Artillery. Containing some fifty pounds of black rifle powder, this mine so impressed Confederate authorities in Richmond that Singer and his associates were immediately withdrawn from their artillery unit stationed along the windswept Texas coast and transferred to the Engineering Department headquartered at the capital in Richmond.

During the two years that the Singer group operated under the direction of the Confederate government, several Union gunboats, troop transports, supply trains, and even the famous ironclad monitor *Tecumseh* fell prey to the group's inventions and were either sent to the bottom or destroyed with great loss of life. The Singer group built at least seven types of underwater contact torpedoes, railroad mines, torpedo boats, and even the famous Confederate submarine *H. L. Hunley*.

Singer agents, operatives, and engineers, directed from their headquarters at the Spotswood Hotel in Richmond, established bases and secret workshops in every major coastal city in the South, from Richmond, Virginia, to Galveston, Texas. Through the deployment of nearly a hundred of their underwater contact mines, they were responsible for thwarting the Federal capture of the last railroad bridge that linked Richmond with the rest of the Confederacy. Some of these mines, which had sent two Federal gunboats to the bottom, were dredged up by Union patrol boats. The design so impressed Union naval officers that they recommended construction of one hundred torpedoes of the same type for use against rebel shipping. John R. Fretwell, Singer's second in command and also a Texan who was assigned the task of protecting the railroad bridge, served for a time on Gen. Robert E. Lee's staff and is presumed to have often briefed the general about the covert activities of the Singer Secret Service Corps.

The deployment of virtually all the group's inventions was conducted at night, often within close proximity to enemy sentries and patrol boats. Many northern newspaper editorials from the period stated that inventions such as those manufactured, deployed, and maintained by the Singer group were nothing less than "infernal machines" and called the entire business of underwater mines and submarine boats "assassination in its worse type" and "unchristian warfare."

From personal meetings conducted with President Jefferson Davis and the secretaries of war, state, and the navy, Singer's agents and inventors took on assignments and operations that have never before been revealed because most Secret Service files that could be collected at Richmond were burned at the end of the war to keep the names of agents and covert operations secret from vengeful parties in the North. Federal authorities constantly hunted members of Singer's group as spies and saboteurs, which prompted Adm. David Porter (commanding the Mississippi Squadron) to sign an order stating that E. C. Singer and his associates were "rebels engaged as agents for the Confederate Government employed for the purpose of furthering the views of said Government in destroying Union vessels by torpedoes and other inventions." In closing, Admiral Porter ordered that if any of Singer's agents or operatives were caught planting or possessing "these inventions," they should be "shot on the spot."

From their torpedo factory located near Galveston, Singer's machinists and craftsmen manufactured and shipped Singer mines and explosive devices throughout Texas, Arkansas, and Louisiana following the Federal capture of the entire Mississippi River during the summer of 1863. With Texas,

Louisiana, Arkansas, and the Indian Territories cut off from the rest of the Confederate states, Singer operatives devised several unique strategies to foil the Union advance into territories of the western Confederacy. David Bradbury, a Singer operative, devised a bold operation to mine the approaches to Fort Esperanza, near Port Lavaca, with modified Singer torpedoes rigged to explode when long lanyards buried with the mines were pulled.

While Singer and most of the organization's personnel carried out covert operations east of the Mississippi, Bradbury and others invented several destructive weapons with the financial backing of the Military Department of the Trans-Mississippi and its commanding general, E. Kirby Smith. One of the most secret inventions was the apparent construction of a second submarine at Houston, Texas, in the final months of the war. From testimony gathered by a Union spy, a talkative Singer operative briefed the spy about the submarine, providing ample proof that the Singer group may very well have constructed a second submarine very similar to the *Hunley*.

Singer and several of his operatives are known to have met frequently with Confederate congressional leaders and officials from the Engineering Department to present various inventions that the group had devised for government consideration. The Confederate government funded a huge ironclad torpedo boat designed soon after Captain Singer and his associates had launched their now famous submarine boat *H. L. Hunley* in Charleston, South Carolina. A Singer operative, and part owner of the *Hunley*, outlined in a letter to Gen. John B. Magruder (military commander of the Department of Texas, New Mexico, and Arizona) one of the boldest projects undertaken by the Singer Secret Service Corps during the war years:

> After considerable hesitation, we were finally ordered by the Secretary of War, to construct one boat at Selma, Alabama, and one at Wilmington, N.C., of the following dimensions. 160 feet long, 28 foot beam and 11 foot hold with flat deck, carrying all their machinery below—to be iron sheathed and with no capacity for guns, and only showing 2 feet above water when ready for work. They are to be arranged with torpedoes, worked from below decks, and through tubes, forward, Aft, and on both sides. It is believed by Engineers of the highest rank, after a full investigation of our plans, that these boats will be perfectly able to raise the Blockade of all the Harbors in the Confederacy.

If this was indeed the case, the Confederacy may very well have developed the first self-propelled torpedo, an accomplishment that could be considered a nautical first that has never before been recorded. Ironclad torpedo vessels to utilize this weapon were constructed in Wilmington, North Carolina; Selma, Alabama; and two in Galveston. Although little is known of the iron-sheathed monsters built east of the Mississippi, a recently discovered reservoir of Confederate documentation regarding the Houston torpedo boats built by the Singer group show that the Texas military highly valued these vessels following the collapse of the eastern Confederacy. Official mention of the vessels can be found on the final page containing the last orders issued by the Gen. Kirby Smith on May 22, 1865.

Throughout the occupied South of 1864 and 1865, Union commanders received general orders including the names and descriptions of Singer and several of his associates and stating that "these desperadoes be apprehended if possible and immediately shot or otherwise speedily got rid of." Some descriptions listed such personal traits and mannerisms as "sloop shouldered steamboat engineer about 140 lbs." and "intelligent Irishman with a large head, high forehead and a florid complexion." One of the more humorous descriptions stated simply that the man is "spitting considerably when he talked."

While Union commanders distributed the names and descriptions of known Singer operatives throughout their respective theaters of operations, the co-designer of the then-lost submarine *H. L. Hunley* was vigorously petitioning Jefferson Davis and the Confederate War Department for government funds to purchase an electric motor. From letters written by agent Baxter Watson comes proof that the Singer group was considering the construction of an electrically powered submarine to replace the *Hunley* after its loss outside Charleston Harbor during the winter of 1864. In communications directed to both the Confederate president and Gen. P. G. T. Beauregard (commander of the Department of South Carolina, Georgia, and Florida), Baxter Watson outlined his daring plan to journey to New York or Washington City to procure "an electromagnetic engine" (risking possible death if captured) for a revolutionary new submarine boat that truly would have been decades ahead of its time.

A brief excerpt from Watson's January 6, 1865, letter to General Beauregard reads: "I propose to build another [submarine] provided I can get the necessary assistance to do so. It will require an electromagnetic engine to propel a boat of that description, as a boat of that kind is impracticable with any other kind of power. I firmly believe that I can destroy the blockade in Charleston in a short time if I get the assistance." Recently discovered documentation makes clear that agents James McClintock and Baxter Watson made extensive experiments with electricity during 1863 for the sole purpose of harnessing its power to propel a small attack submarine.

Of the various Singer operatives known of and pursued by Federal authorities, none was more notorious than Henry Dillingham, Confederate agent, saboteur, and friend of Lt. George E. Dixon (final commander of the *Hunley*). A steamboat engineer hailing from Mobile, Alabama, Dillingham had served as a crewman/engineer aboard the *Hunley* for several weeks during the fall of 1863. In the closing months of the war Dillingham (under the direction of President Jefferson Davis) led a group of saboteurs deep into occupied Kentucky and Missouri with orders to destroy an important Federal railroad bridge and any Union transport vessels that came within their reach.

Among the group's secret weaponry and explosive devices could well have been the mysterious coal torpedo designed by Missouri native Thomas Courtenay in late 1863. Disguised as a large lump of coal and containing some ten pounds of black powder, this device had been used successfully to destroy the Union vessel USS *Greyhound* on the James River, Virginia, and was soon a source of constant dread to Union naval officers. If a rebel spy was successful in planting one of these infernal devices in a Union coal bunker, it was almost impossible to detect, and once the mine had been innocently shoveled onto the fires beneath a ship's boiler, the vessel's fate was sealed. Jefferson Davis is known to have been impressed with Courtenay's coal torpedo, and it is possible that he may well have outfitted Dillingham and his agents with several of these devices before their departure from Richmond.

Some sources state as fact that one of these devices was responsible for destroying the Union steam transport *Sultana* in late April 1865 as it ferried ex-prisoners of war north on the Mississippi River (the same area assigned to Henry Dillingham and his agents). The sinking of the *Sultana*, following the unexplained explosion of its steam boilers, is still regarded as the worst nautical disaster to have taken place in American waters. In an attempt to hang the assassination of Abraham Lincoln around the necks of the Confederate Secret Service, Federal prosecutors singled out Henry Dillingham and several of his comrades for their behind-the-lines activities, and testimony concerning their escapades and sabotage to Union river transports and supply depots during the final weeks of the war fill several pages of testimony taken at the Lincoln assassination trial.

While testimony concerning the activities of the Confederate Secret Service was being gathered in Washington during late April 1865, Singer agents and operatives continued to carry out their varied assignments until virtually the last Confederate guns fell silent. With the collapse of Richmond, Henry Leovy (a Singer associate and leader of a counterintelligence bureau who owned a full third of the *Hunley*), joined Jefferson Davis and his cabinet in their escape southward. When the group disbanded in rural Georgia some weeks later, Leovy (a prewar friend of Singer agent Horace L. Hunley, for whom the famous Confederate submarine was named), took on the guise of a French interpreter and accompanied the defrocked Confederate secretary of state, Judah Benjamin, on his escape to Florida.

The outlandish and far-fetched wartime activities of the Singer Secret Service Corps were very fragmented and practically unknown prior to the establishment (at the Warren Lasch Conservation Center in North Charleston, South Carolina) of a comprehensive archive of materials relating to the recently raised and excavated Confederate submarine *H. L. Hunley*. The backgrounds of all investors and individuals associated with the project had to be studied in depth to locate all contemporary documentation relative to this first successful combat submarine. This extensive investigation was vital to establish, with some confidence, the sequence of events that led to the *Hunley*'s final construction, and eventual success, off Charleston Harbor on the night of February 17, 1864.

During this exhaustive research into the various backgrounds of the personnel associated with E. C. Singer and his torpedo engineers, activities other than the manufacture and deployment of underwater contact mines and the building of the *Hunley* came to light. Previously unknown Confederate documentation began to reveal itself, indicating that Singer and those associated with his Secret Service Corps (an unknown name prior to the commencement of this investigation) were involved in activities far more covert and secret than anyone had previously guessed.

During gathering of documentation from archival repositories throughout Washington, D.C.; Mobile, Alabama; and several of the former Confederate states, information presented itself that proved beyond a doubt that historians had overlooked an important chapter of the American Civil War. The adventures and operations of the men associated with Edgar Singer, his torpedo operations, and the nine-man *Hunley* had until now been fragmented and practically unknown, and only through the painstaking research conducted in the past few years has the true history and extent of their wartime activities been realized.

The recent discovery, recovery, and excavation of the *Hunley* have elevated interest in underwater warfare conducted during the Civil War to a new height. With various *Hunley* documentaries and even a movie (Turner Network Television, 1999) having been produced since the discovery, there can be little doubt that the world has finally taken notice of the engineers and personnel associated with this small, ill-fated Confederate submarine that was first in history to sink an enemy ship in wartime. Soon after the recovery, international media attention was directed toward the place of restoration at the Warren Lasch Conservation Center, and updates regarding artifacts found within the hull during the excavation appeared on all the major networks (and tickets to view the history-making vessel sold out within hours whenever offered).

With such a rich, previously untold story swirling around the *Hunley* and those who built, deployed, and ultimately died in the vessel, I decided to record these various exploits and recently discovered covert missions shared by those southern engineers and adventurers so they would not be lost to history. The work may seem to have a somewhat southern slant; this is perhaps true, but considering the fact that we will be discussing the history of a Confederate Secret Service company and the incredible adventures shared by its colorful members, this "slant" cannot be helped. I hope that the reader finds this book worthy of the brave men written of within.

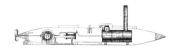

I Formation and Deployment

THE WINDSWEPT COAST OF SOUTHERN TEXAS, where Edgar C. Singer and several future members of his Secret Service Corps were then stationed during the Civil War, was about as far removed from Richmond, Virginia, and the war in the east as one could get yet still be within the borders of the Confederacy. Texas, unlike other states in the Union, had been an independent republic less than a generation before the South declared independence, which perhaps explains why Texans vigorously supported the ideas of states' rights and secession when the United States fragmented in early 1861. At the Texas Secession Convention held at the state capital in Austin in late January, several delegates openly expressed the opinion that Texas should reestablish itself as an independent republic and leave the other six recently seceded southern states to form their own political alliances.[1]

On February 1, 1861, to the great dismay of the hero of Texas independence, Sam Houston, an overwhelming majority of delegates signed the Texas Ordinance of Secession and hastily arranged a statewide vote for February 23 that would either ratify or reject their decision (Texas was the only state to put the question of secession before voters). To no one's surprise the citizens of the Lone Star State quickly ratified the decisions made by their delegates (in Galveston the vote was 765 to 33) and immediately sent representatives to form alliances with the other states that had already left the Union.

Unlike other southern states that had been settled for generations, Texas in many ways was still considered a frontier. Galveston, the second-largest city and main trading port, was incorporated just twenty-three years earlier, so this southwestern state was still undergoing important changes at the time of secession from the United States. Before the Civil War Galveston had become a well-established gateway to European immigrants seeking a new way of life and cheap land. According to the 1860 census, almost 40 percent of the population of Galveston was born elsewhere, the vast majority having emigrated from Germany.[2]

Singer, like almost everyone else then residing in Texas, was himself an immigrant who had come to the state with his wife and family in the late 1840s.[3] Born in Ohio on October 21, 1825, the young Singer was no stranger to new ideas and inventions, for at an early age his uncle, Isaac Merritt Singer, had designed and built the first sewing machine in the home of Edgar's father at Lockport, Illinois.[4] Settling in the small coastal town of Port Lavaca, located some ninety miles south of Galveston, Edgar Singer worked as a local gunsmith.[5] Standing just over six feet three inches, the burly Texan towered over his contemporaries, which perhaps made him slightly intimidating.[6]

At the start of hostilities in the spring of 1861, Singer, then thirty-six years old, joined Capt. Daniel D. Shea's company of Texas Light Artillery stationed on the shores of Matagorda Bay near his home of Port Lavaca.[7] During June and July, thousands of young Texans swarmed to Confederate recruiting stations throughout the state, joined newly formed regiments with patriotic names, and hurried eastward in an almost carnival atmosphere to take part in a war all thought would be decided by Christmas. After Federal defeats at Manassas, Virginia, in July and Wilson's Creek, Missouri, in August, southerners looked favorably on a swift end to the conflict and were soon shocked and surprised at the enemy's willingness to carry the war to a military decision.

When hostilities dragged on in the eastern theater and because so many young fighting men were then absent, Texans turned a wary eye toward coastal regions, and rumors of imminent Yankee invasion from the sea started circulating throughout the state. The situation in the fall of 1861 can best be summed up from a letter written by Texas general Paul Hébert to the Confederate secretary of war on September 27 in which he stated that he found the coastal defenses "in almost a defenseless state, and in almost total want of proper works and armaments; the task of defending successfully any point against an attack of any magnitude amounts to a military impossibility."[8]

To complicate matters, the Confederate secretary of war issued orders to General Hébert (then military commander of the Department of Texas) on March 14, 1862, directing him to dispatch all available troops except those "necessary to man the coast batteries" to Maj. Gen. Earl Van Dorn, then at Little Rock, Arkansas.[9] The South had suffered triple defeats in February with the loss of Forts Henry and Donelson in Tennessee, and the state capital at Nashville had to be hastily evacuated, so Gen. Sidney Johnston, then attempting to rally troops in northern Mississippi for a counterassault, needed every man who could shoulder a musket.[10] With every possible soldier desperately needed both east and west of the Mississippi River, prospects for increased coastal reinforcements for Texas evaporated.

To increase efficiency in the western regions of the Gulf of Mexico, the US Navy split its blockading forces, then operating in that theater, into separate eastern and western components. David G. Farragut was quickly given command of the West Gulf Blockading Squadron and assumed the monumental task of sealing off such established Confederate port cities as

Mobile, New Orleans, and Galveston. With a naval service record stretching back to the War of 1812, the southern-born Farragut, then in his early sixties, was fiercely loyal to the Union and, unlike many naval officers with a southern heritage, opted to stay with the flag when the country divided. When some former naval comrades who had recently resigned approached him in an effort to recruit his services for the southern cause, Farragut responded, "Mind what I tell you: You fellows will catch the devil before you get through with this business."[11]

With Mobile and New Orleans well defended, Farragut took over his new command with an eye toward operations against the Texas coast, where defenses were known to be undermanned, inadequate, and scattered. The Union War Department hoped that the reoccupation of Galveston would strike a blow to the morale of the still swaggering rebels who had yet to lose a major city or port. In early March 1862, Farragut stated in a letter to a subordinate, "Galveston will be looked to at my earliest convenience."[12]

Farragut apparently intended to first assault the Texas coast after taking command of his squadron, but the capture of New Orleans, which had vast quantities of southern cotton piled on its wharves, soon overshadowed the intended attack on Galveston. While Farragut and his officers planned a Federal naval assault on New Orleans from the Gulf of Mexico, some future agents of the Singer Secret Service Corps (then residing there) were secretly experimenting with a small submarine vessel that the Confederate government had recently granted a privateering commission. Christened the *Pioneer*, this forerunner of the now-famous Singer-built submarine *H. L. Hunley* was the creation of riverboat pilot James R. McClintock; attorney, wealthy landowner, and New Orleans customs inspector Horace L. Hunley; steam gauge manufacturer Baxter Watson; and editor and part owner of the *New Orleans Daily Delta* newspaper Henry J. Leovy.[13]

All four men would later be intimately involved with the design, manufacture, and operations of the *Hunley*, as well as other government-sanctioned operations vital to the southern war effort. Confederate Secretary of State Judah P. Benjamin (responsible for both overseas relations and dispersal of Secret Service funds) was the prewar friend of Henry Leovy; thus, the four men's partnership was well connected with at least one important official in the Confederate government.[14]

After the war James McClintock wrote to Matthew Maury (the Confederate "Chief of Sea-Coast, Harbor and River Defenses of the South"), describing this early war predecessor of the Singer Secret Service Corps' most ambitious project.

> In the years 1861, 62, and 63, I in connection with others [undoubtedly those associated with him in the Singer Secret Service Corps] was engaged in invent-

ing and constructing a submarine boat or boat for running under the water at any required depth from the surface. At New Orleans in 1862 we built the first boat, she was made of iron ¼ inch thick. The boat was of a cigar shape 30 feet long and 4 feet in diameter. This boat demonstrated to us the fact that we could construct a boat that would move at will in any direction desired, and at any distance from the surface. As we were unable to see objects after passing under the water, the boat was steered by a compass, which at times acted so slow, that the boat would at times alter her course for one or two minutes before it would be discovered, thus losing the direct course and so compel the operator to come to the top of the water more frequently than he otherwise would.[15]

In spite of the *Pioneer*'s apparent shortcomings, the Confederate State Department issued a letter of marque (privateer commission) on March 31, 1862, bestowing on the *Pioneer* the international distinction of being the world's first and only submarine privateer. Its only offensive weapon was described as simply a "magazine of powder,"[16] so it appears that the designers had been independently experimenting with underwater explosives early in the war.

With an intended occupying force of some eighteen thousand Union troops under the command of Gen. Benjamin Butler in reserve, Farragut's fleet steamed up the Mississippi River and opened fire on the two rebel forts that barred his passage to New Orleans with mortar boats on the morning of April 18. After a small section of the river had been opened, Farragut pushed his fleet past the two Confederate strongholds on the night of April 23.

By the early morning on April 25, the partially battered Union fleet approached the last obstacle that hindered the capture of the large rebel commerce center, the Confederate batteries at Chalmette just four miles below New Orleans. After a brief exchange of fire with the vastly superior Union armada, the southern cannons were swiftly silenced, and Farragut and his attacking ships slowly steamed upriver toward the defenseless port. As the enemy fleet approached, retreating Confederates put the torch to more than thirty thousand bales of cotton shortly after noon, and within an hour the entire river front was ablaze. For the next three days heated negotiations between city hall and representatives of Farragut's fleet continued until the city officially surrendered on April 29.[17]

The sudden loss of the cotton-rich port city of New Orleans sent shock waves through the Confederacy, for not only was a valuable munitions manufacturing facility and supply center lost but access to the southern Mississippi River was now cut off. With their army in retreat the submarine partners McClintock, Leovy, Hunley, and Watson hastily sunk their invention in the New Basin Canal near the Custom House and fled to Mobile to continue their experiments.

After the loss of New Orleans, Confederate forces had been driven from both ends of the Mississippi River, and only the middle sections from just north of Memphis to Baton Rouge still remained in southern hands.[18] As General Butler set up his headquarters and assessed the damage to the dock area prior to the rebel army's departure, Flag Officer David G. Farragut was writing to the commander of his small blockading squadron stationed off the Texas coast. Because New Orleans was in Federal hands, Farragut thought that Confederate forces defending the vulnerable city of Galveston might be persuaded to abandon their positions, and he wrote a short communication to that effect to the head of the Texas squadron, Capt. Henry Eagle of the USS *Santee*: "Demand the surrender of Galveston. Tell them it is only a matter of time, and I will be along your way soon."[19]

To comply with this order, Captain Eagle sent a proclamation to the military defenders of Galveston on May 17, 1862: "In a few days the naval and land forces of the United States Government will appear off the town of Galveston to enforce its surrender. To prevent the effusion of blood and destruction of property which would result from the bombardment of your town I hereby demand the surrender of the place, with its fortifications and all batteries in its vicinity, with all arms and munitions of war. I trust you will comply with this humane demand."[20]

The civilian population of the city panicked at this new threat from the Federal blockaders, and the roads leading to Houston were soon clogged with carriages and refugees heading inland. The pledge to shell the city brought commerce to a standstill, and in the anxious weeks that followed (the Union threat did not materialize for several months) the streets of the once busy port were practically empty.

The events then taking place at Galveston, some ninety miles north of Port Lavaca, were well known to Edgar C. Singer and the several men within his artillery company that would soon make up the inner circle of his Secret Service Corps. Since the early months of the war Shea's Battalion of Texas Artillery had established several crude defensive positions and gun emplacements around Port Lavaca and nearby Matagorda Island. With just 180 men and four 24-pounders, two 12-pounders, and one 6-pounder at his disposal, Captain Shea's command was in no position to defend itself against a determined Federal attack. Except for sixteen rounds being fired at a passing Federal blockader on December 14, 1861,[21] his men had not seen action, and with no infantry support to speak of (except a small unit of Port Lavaca and Indianola Home Guard) the odds of holding out against any Union attack were bleak.

To remedy this situation, Shea's artillery company began construction of Fort Esperanza in early 1862 at the northern tip of Matagorda Island overlooking the narrow inlet at Pass Cavallo, and after several months the fort became the main garrison for the company. Nearby Port Lavaca and the harbor town of Indianola were by no means important military targets, and Fort Esperanza and the surrounding sandwork fortifications, armed primarily with "Quaker guns" (large logs painted black and fabricated to look like gun emplacements), were little more than a backwater garrison on a windswept, mosquito-infested strip of sand some ninety miles south of Galveston.

Through the late spring and summer of 1862, the poorly equipped men of Shea's isolated garrison scanned the empty horizon of the Gulf of Mexico searching for the expected Federal invasion armada. Union raiding parties occasionally came ashore on the outer islands and established temporary outposts, so Texas citizens scattered in small coastal towns lived in constant dread of attack. Meanwhile, all coastal garrisons that summer suffered from epidemics of camp measles and yellow fever.[22]

Because the bulk of Texas regiments had been deployed to the eastern regions of the Confederacy, Adm. David G. Farragut turned a serious eye toward the city of Galveston in mid-September and issued orders to Capt. William Renshaw to take the rebel port if possible: "You will proceed down the coast of Texas with the other vessels, keeping a good lookout for vessels running the blockade, and whenever you think you can enter the sounds on the coast and destroy the temporary defenses, you will do so and gain command of the inland navigation. Galveston appears to be the port most likely for you to be able to enter, if the forts are not too formidable."[23]

Capt. Henry Eagle of the USS *Santee* had predicted several months earlier how the capitulation of Galveston could be assured when he reported that the city would probably surrender if confronted by mortar boats, such as those used in the bombardment of the forts below New Orleans in April 1862. With this in mind Farragut transferred several such vessels to the Texas coast, and at about 6 a.m. on the morning of October 4, 1862, this small but formidable Federal fleet anchored off Galveston and immediately demanded its surrender.

At this serious threat to the city, Confederate military officials, under a flag of truce, negotiated a four-day grace period with the commander of the attacking armada and issued the following proclamation to the citizens of the isolated city: "Notice! Headquarters, Galveston, October 4th, 1862, 10 o'clock P.M. The Commander of the Federal Naval fleet having granted four days time to remove the women and children from the city, notice is hereby given to the citizens, that they may avail themselves of the opportunity of leaving."[24]

The Union warships USS *Harriet Lane*, *Owasco*, *Clifton*, and *Westfield* were in plain sight of the astonished citizens of Galveston. A general retreat toward Houston began without delay, while military units hastily packed up or destroyed any ordnance that might be of value to the enemy. From their ships anchored offshore Union naval officers could see the Confederates removing or dismantling their fortifications and quickly sent a strong protest to Colonel Cook (military commander of Galveston) stating that their grace period applied only to the removal of noncombatants, not the redeployment of ordnance. In reply Colonel Cook bluffed that the city was undergoing an epidemic of yellow fever and invited them to come ashore and see for themselves.[25] Union naval officers reluctantly decided not to interfere with the rebel retreat and casually bided their time until the morning of October 8, when the city was officially surrendered.[26]

The surrender of Galveston resulted in a general panic throughout the state. Governor Francis Lubbock and his staff immediately departed from Austin and hastened to the confused theater of operations. Upon arriving within the disorganized Confederate lines that had been hastily drawn up near the city, Lubbock fumed at his officers for not burning Galveston before surrendering it and openly voiced his concerns that no naval forces of any kind were available to scout enemy positions or guard against troop movements. With the valuable port city now in enemy hands, the panic-stricken citizens in the coastal settlements thought that it was only a matter of time before all of Texas would be under the yoke of the powerful northern invader.[27]

At Port Lavaca, Pvt. Edgar C. Singer, and the rest of the men attached to Shea's Texas Artillery, quickly assessed the situation and with the aid of the Home Guard units, hurriedly reinforced their positions in preparation for an enemy attack expected within days. In the weeks that followed, Major Shea oversaw the renewed construction of his defensive works and sent scouting parties up and down the coast to watch for any sign of the expected foe. They would not have to wait long.[28]

On October 31, 1862, a small Union naval force steamed unopposed into Matagorda Bay and commenced shelling the small coastal settlement of Port Lavaca. While Federal warships dropped exploding shells on the streets and very homes of the men who crewed the defending guns, one can only imagine the tenacity with which Shea's soldiers and their Home Guard support units fought to drive off the attackers. For two days Union guns shelled the isolated port, firing over 250 rounds into the town and causing destruction on a scale heretofore not seen by either its defenders or inhabitants.

A Confederate report on the engagement included the following information:

> Headquarters, Lavaca, Texas, November 1, 1862. Sir: By order of Major Daniel D. Shea, commanding this post, I have the honor to make, for the information of the general commanding this district, the following report of an engagement between the federal steamers and the batteries at this point: On

the morning of October 31 two federal steamers appeared in sight, evidently steering for this place. . . . About 1 p.m. they sent a boat with a flag of truce on shore, which was met by Major Shea, accompanied by four of the citizens of the town . . . [and] a demand was made for the surrender of the town. They were answered by the commanding officer that he was there to defend it, and . . . a demand was then made for time to remove the women, children and sick persons from town.

The officer in charge of the flag replied that . . . in consideration of the fact that an epidemic [yellow fever] was still raging in the town he would extend the time to one hour and a half; at the expiration of which period they moved up abreast the town and opened fire from both steamers upon both the town and batteries. . . . Our batteries promptly returned the fire. . . . The steamers were struck several times and one of them partially disabled, as they immediately steamed off out of range of our batteries, where they again cast anchor and kept up a steady fire upon the town and batteries until night shut in.

On the next morning, November 1, they again opened fire upon the town and batteries. . . . No lives were lost on our side, but the enemy succeeded in doing considerable damage to the town, tearing up the streets and riddling the houses and otherwise damaging the place. The enemy fired in all 252 shot and shell, 174 the first day and 78 the second, nearly all of them from 32 and 64 pounder rifled guns. George E. Conklin, Lieutenant and Adjutant.[29]

The unopposed movements of Union ships and their numerous landing parties around the once-secure region of Matagorda Bay had a life-changing effect on Singer and several of his friends, for it was during this period of crisis and uncertainty that Singer is said to have first considered the feasibility of underwater warfare as a means of defense against similar attacks. With the aid of several men in his company, Singer began experimenting with small charges of gunpowder in a water-filled barrel behind his house in Port Lavaca and soon became confident about the destructive power that could be unleashed.[30]

Singer and his associates would soon become familiar with underwater explosives, and the siege of his hometown prompted them to engage in the business of torpedo warfare, a clandestine occupation that would soon bring great destruction to the enemy on both land and sea.[31] When they exhausted their supply of powder, Singer immediately approached his commanding officer, explained his proposed system of underwater defense, and requested several additional pounds of gunpowder to continue tests in a nearby bay. At first Major Shea was skeptical of Singer's idea and refused to release any of his unit's valuable gunpowder because enemy warships were patrolling Matagorda Bay and Federal landing parties were terrorizing outlying settlements.

During this early experimentation with underwater explosives Edgar Singer formed a partnership with Port Lavaca physician John R. Fretwell, a forty-seven-year-old native of Mississippi and head of the local Masonic Lodge, of which Singer himself was a member. With a prewar worth estimated to have been over fifty thousand dollars, Fretwell would have been considered a wealthy man for the times, and some of that wealth might have been used to fund some of Singer's early experiments. In late April 1861, Fretwell enlisted as a private in the Lavaca Guards under Capt. A. H. Phillips and was soon promoted to second lieutenant. Originally organized as a light infantry company, the Lavaca Guards were soon absorbed into Captain Shea's artillery company stationed near the city.[32] However, official company records on file at the National Archives list him with the rank of private.

In the early days of the conflict Fretwell had demonstrated that he was an ingenious inventor and had the distinction of fabricating and casting the first two Confederate cannons made in Texas. A local historian wrote about Fretwell after the war: "He cast two iron pieces, and presented them to Lavaca County in the early part of 1861."[33] Fretwell most likely financed the cannons as well, and they may well have formed the nucleus of Captain Shea's available ordnance.

The fall of Galveston caused a great panic up and down the Texas coast throughout October and early November 1862. Major Shea's artillery batteries scattered around Matagorda Bay were powerless to stop Union vessels from entering, and there was a fear that Federal regiments, then garrisoned in occupied New Orleans, might be sent to invade Texas. Orders were issued from Houston to immediately obstruct the mouth of every Texas waterway deep enough to allow an enemy transport vessel to enter.[34] Consequently, Major Shea reconsidered Singer's request for additional gunpowder. In a postwar interview Singer stated that "he persisted and finally obtained eight pounds of powder and offered to demonstrate his invention."[35]

Since every vessel throughout the region that could float was then in state service, an old partially beached hulk that had been a Port Lavaca nautical hazard for years was chosen as the target to demonstrate the effectiveness of a submerged explosive.[36] As the day for the demonstration neared, Singer and Fretwell presumably tested and retested the small metal container that would hold the submerged eight pounds of gunpowder, making sure that not a single drop of water could enter.

After the waterproofing tests were completed, the men carefully filled their underwater bomb with gunpowder, set the ingenious spring-loaded detonating rod, and anchored the device underwater alongside the wreck. "The mine was placed alongside and set off, and the vessel was blown to atoms. Commander Shea was astounded and immediately ordered Singer to report to General Magruder at Houston."[37]

Gen. John B. Magruder, "Prince John" to his old army friends and veterans of the spring and summer campaigns in Virginia, arrived in Houston to assume command of the Department of Texas, New Mexico, and Arizona on November 29, 1862 (replacing the unpopular Gen. Paul Hébert, who had surrendered Galveston without a fight the month before), so Singer and his associates were not dispatched to Houston until perhaps sometime in December.[38]

With General Magruder's arrival in Houston, the fading morale of the Texas military was recharged, for although the general had a reputation as a drinker, he was also well known as a fighter. Thomas J. "Stonewall" Jackson had served under Magruder in the Mexican War and would later write of his old commander: "If any fighting was to be done, Magruder would be on hand."[39]

At the time "Prince John" relieved Gen. Paul Hébert, Governor Lubbock was openly calling for the recapture of Galveston and wasted little time in letting Magruder know this. With all available Confederate forces then massed near the sparsely garrisoned port, Magruder and his staff pored over regional maps to locate a suitable avenue of attack. At about this time Singer and Fretwell arrived in Houston with diagrams and sketches of their new torpedo system. However, Magruder and his staff may well have been too busy to give the matter of underwater warfare much thought. Unfortunately, extremely little is known of the events surrounding Singer's journey to Confederate military headquarters, and his proposed demonstrations may have been placed on hold until after the attack to retake Galveston had been attempted.

Just under ten thousand troops were then present within Texas, and of those many were tied to established coastal fortifications and could not be utilized in the forthcoming operation.[40] Magruder wrote letters with pleas for men and arms to both his superiors in Little Rock, Arkansas, and to Governor Lubbock in Austin. With a small number of lightly armed steamers called "cottonclads" (because of protective barriers of cotton bales piled on their decks) at his disposal, Magruder and his staff decided to attack the newly arrived Galveston garrison with a combined land and naval assault.

Some three hundred troopers from the 5th and 7th Texas Cavalry Regiments were detached from their land-based units and placed aboard the cottonclads as sharpshooters and boarding parties. In the aftermath of the battle these cavalrymen came to be known throughout Texas as the Galveston "Horse Marines" or "Horsemen of the Sea."[41] With his crude naval flotilla reinforced and several batteries of heavy artillery poised to pound the Federal garrison into submission, General Magruder set the date for his daring attack on Galveston for December 31, 1862.

The joint land and sea attack launched against Galveston just after dusk has become something of a Texas legend, and stories singing the praises of the "Horsemen of the Sea" and the others who took part in the engagement have reached mythic proportions. The small Union fleet anchored in the harbor and the garrison troops encamped along the wharves were apparently caught completely by surprise. The battle itself was one of the most unique engagements ever fought in North America, for practically the entire operation transpired between the hours of 2:00 and 6:00 a.m. on New Year's Day. Simultaneous naval and land assaults were launched within sight of one another, while Confederate boarding parties swarmed from their crudely protected cottonclads with shotguns and cutlasses onto the decks of Federal warships in the darkness (one of the few instances of capture by boarding recorded during the Civil War).[42]

The surrender of both the Galveston blockading squadron and garrison troops marked the only time during the Civil War that the South would recapture a major port city. In the battle's aftermath General Magruder basked in the attention showered on him by the southern press, joyful Texas citizens, and a grateful Confederate Congress, who had hastened to pass a special resolution praising Magruder and his forces for their brilliant victory. In Houston both a ball and parade were staged in his honor, and the thankful citizens of Texas had taken up a collection to present him with an ornate dress saber. This in all likelihood was the state of affairs that greeted Edgar Singer and John Fretwell upon their arrival in Houston.[43]

Singer and Fretwell in all probability were forced to cool their heels in a Houston hotel until General Magruder could find a suitable opportunity to receive them, which was probably early in January 1863. Singer stated in an interview that "General Magruder was skeptical of the invention," but he gave Singer twenty-five pounds of powder to conduct an experiment in Buffalo Bayou.[44] Singer and Fretwell decided to stage a grand demonstration.

General Magruder was well aware of the inadequacy of his land forces and knew better than anyone else that Galveston would again be taken by the Federals unless more men were forthcoming or a new system of water defense (other than simple obstructions) was incorporated. Magruder may well have concluded that mine warfare might be the answer to defending the vulnerable Texas coast against an invasion from the sea. The following account of the Houston torpedo demonstration appeared in the *San Antonio Express*: "An old scow lay just above Houston; the mine was submerged and the scow floated down over it. General Magruder, his staff and many citizens of Houston assembled on the banks of the bayou were warned to stand back, so that their lives may not be endangered by flying timbers. When the boat struck the mine she was blown into kindling wood and the observers frantically sought trees for protection."[45]

Soon after the old scow "was blown into kindling," General Magruder and his staff arranged a meeting with the two South Texas inventors and questioned them about the hazards involved in utilizing this new explosive device for the defense of Galveston. After assurance that their torpedo was relatively safe in the hands of an experienced operator, the general sent the men back to their company immediately to begin fabrication of six torpedoes for use in Galveston Bay.[46]

Upon arrival in Port Lavaca, Singer was immediately granted a leave of absence to fabricate the explosive devices in his small workshop. Because Magruder was so impressed with Singer's invention, Major Shea and several other officers immediately placed orders for several torpedoes of their own. Singer's unique organization most likely was formed during early January 1863, as the first underwater mines were deployed in Texas waters by the end of the month.[47]

The torpedo itself was a rather simple device consisting of little more than a cylinder-shaped, watertight, metal canister of gunpowder and a spring-loaded detonating rod. A narrow metal pin with four prongs was inserted through a small hole in the shaft of the spring-loaded plunger to trigger and explode the submerged torpedo (see diagram of the "percussion torpedo"). When a passing object jarred this triggering pin free from its position, the rod slammed into the end of the metal canister and detonated two internal percussion caps, which in turn exploded the powder. The entire device was chained to a small anchor and submerged about three feet beneath the surface of the water with the detonating trigger in an upright position so it could be jerked free when a ship's hull brushed past.[48]

The anchoring and deployment of a Singer torpedo was somewhat tricky and required several men to set the triggering mechanism once the device was anchored in place. The deployment vessels often carried several loaded torpedoes, and because these hazardous manipulations often had to be executed under cover of darkness, the men involved were under constant danger of blowing themselves up. The operator armed the mine by pulling back the spring-loaded detonating rod and placing a metal safety pin through a small hole. He then attached this safety pin to a fifty-foot cord that remained in the boat after the mine was submerged in a horizontal position (with the four-pronged detonating trigger pointing up). When the deployment boat had moved away fifty feet, the operator withdrew the safety pin from the detonator, and the mine was then operational.[49]

It was obvious from the start of the operation that several men were necessary to fabricate, maintain, and deploy these torpedoes, and Singer and Fretwell looked no further than the membership roster of their local Masonic Lodge.[50] Most of the soldiers in Shea's company were locals, and many were either unfit or too old for active field service east of the Mississippi River, where the vast majority of Texas regiments were then deployed. Their military duties encompassed little more than occasionally rotating a gun crew and scanning the empty horizon for passing enemy ships, so several jumped at the chance to join Singer and Fretwell in their unique patriotic venture.

We know the names of seventeen operatives assigned to the Singer Secret Service Corps during the war years,[51] but as many as another fifteen or twenty members are unknown because most Secret Service files were burned in the final days of the war to keep past operations and the names of agents secret from Federal authorities. The backgrounds of some members are in fact known from the 1860 census records of Port Lavaca, New Orleans, and Mobile.

In the days before the invention of the polygraph and other techniques to determine an individual's trustworthiness, the safest course in any secret operation was to deal only with people one had known from personal contact. This also reduced the risks of recruiting an informant or spy dispatched by the enemy. A pool of trustworthy individuals could be recruited from classmates, neighbors, family, and fellow members of the same church, club, or lodge for a clandestine operation.[52] Singer and Fretwell recruited their core group from Fretwell's local Masonic Lodge.[53]

One of the first men approached to join their unique venture was James Jones, a thirty-five-year-old Port Lavaca jeweler enlisted in Shea's artillery, a native of Kentucky who had migrated to Texas several years before the South's secession.[54] The unique skills of a jeweler would have been quite valuable in the fabrication of delicate detonating devices, and these talents were apparently considered important enough by both founding partners to extend Jones an offer of membership. David Bradbury, a fifty-one-year-old local contractor, was an early Singer inductee who eventually headed up all operations west of the Mississippi River. Bradbury himself, like virtually all early members, was affiliated with both the Masonic Lodge and Shea's company of Texas artillery.[55]

Another founding member was William Longnecker, the fifty-nine-year-old owner of the Port Lavaca Livery Stable. In spite of Longnecker's advanced age for the 1860s, he was one of the first to be dispatched eastward to plant torpedoes in enemy-patrolled waterways. Local merchant John D. Braman at age twenty-seven was the youngest member of the group during the early days of the operation, and within months was in the besieged city of Charleston, South Carolina, where he eventually became part owner and crewman aboard the soon-to-be-constructed submarine *H. L. Hunley*. Robert W. Dunn, another future *Hunley* crewman who invested fifteen hundred dollars toward the vessel's construction, was a forty-year-old Port Lavaca merchant and fellow Masonic Lodge member who with his wife, Fannie, had migrated to Texas from Kentucky several years before the South's secession.

C. E. Frary, a thirty-year-old Canadian-born Port Lavaca carpenter and

Edgar Singer's brother-in-law, was another future *Hunley* crewman who would soon journey eastward to Richmond to demonstrate the group's newly designed contact torpedo. Of this early core of associates, thirty-seven-year-old merchant B. A. (Gus) Whitney was perhaps one of the more energetic and charismatic members, who journeyed to Charleston to meet with Confederate military officials regarding the deployment of the *Hunley*. Of the founding member/partners that H. N. Hill lists in the *San Antonio Express*, only F. M. Tucker and L. C. Hirshberger do not appear on the 1860 census of Port Lavaca, so no information is known about them.[56]

Once Singer had selected the core group of his torpedo company, Major Shea immediately detached or granted extended furloughs to all members. David Bradbury apparently headed up Singer's Port Lavaca torpedo manufacturing facility, for his name appears on the first known document dealing with the subject, and he was soon placed in charge of distributing Singer-Fretwell torpedoes (as some historians have named them) to all regions west of the Mississippi River.[57]

In the weeks that followed, Singer's group worked around the clock to fill the torpedo orders placed by General Magruder and Major Shea. Some workers used tin snips to cut sheet metal into an appropriate pattern that could be rolled to fabricate the powder compartment, while others manufactured coiled springs or drilled holes in recently machined detonating rods. For safety reasons the torpedoes themselves were never primed with powder while still at the manufacturing facility (gunpowder would be poured only prior to anchoring), so probably little or none was stored on the premises.

Within weeks of their demonstration for Magruder, they delivered the first completed Singer-Fretwell torpedoes to the military defenders of Fort Esperanza. We know that the first Singer mines manufactured were deployed in late January or early February near the fort, according to a brief report to Maj. A. J. Dickinson at Galveston: "Head Quarters Shea's Battalion, Lavaca, Texas, February 2nd, 1863. Major: I have the honor to inform you that the torpedoes have been placed at Pass Cavallo Bar. The enclosed signals [not found with this report] have been adapted to procure a pilot who will be in readiness, should any Confederate Vessel need his assistance or intend to come over the Bar."[58] An additional report also dated February 2 states that "Pass Lavaca," the last water route leading to Port Lavaca, had been mined with Singer torpedoes as well.[59] Perhaps Singer and his men wanted to protect their own property and homes from Federal assault prior to mining the water approaches leading to the recently recaptured port of Galveston.

Within days following the deployment of torpedoes near Fort Esperanza and Port Lavaca, Singer and some of his torpedo men (with members James Jones and William Longnecker known to have been present) were dispatched to Houston with the six torpedoes ordered by General Magruder for use in

Galveston Bay. The following brief note confirms that torpedo manufacture had started in early January: "Port Lavaca, February 5, 1863. I hereby certify that there is due James Jones one hundred and fifteen dollars for one month and three days work at Port Lavaca in the manufacture of Singer's Submarine Batteries for E. C. Singer, R. W. Dunn, William Longnecker and D. Bradbury. And it is the understanding that it will be paid to him upon his arrival in Houston."[60]

The torpedoes were turned over to military authorities upon their arrival in Houston, as confirmed in the Houston Military Headquarters order book in early February: "The torpedoes prepared by Mr. Singer, six in number, will be placed by Colonel Forshey in such position in the water of Galveston Harbor. . . . Col. Forshey will make known the particulars of the position, arrangement and management of these machines to the commanding officer of Galveston, who will give such further orders."[61]

It seems that bold plans had been formulated within the newly established Singer-Fretwell organization since returning from their demonstrations in Houston, as noted in the following communication requesting that James Jones and William Longnecker be redeployed to establish torpedo operations in the eastern theater:

> Headquarters, Artillery Battalion, Lavaca, Texas, February 6, 1863. To Captain E. P. Turner A. General, Houston, Texas, Captain: Permit me to introduce to your favorable notice Messrs. William Longnecker and James Jones. Members of my Battalion, they are engaged in the Torpedo enterprise, and are desirous to travel east, where they can carry out their operations more efficiently. It is necessary that they should be furnished with the proper papers from your office, to protect them as Confederate State Troops in case of capture by the enemy. Captain Bradbury and others are apparently within the limits of my command. I have the honor to be Very Respectfully your obedient servant David Shea Maj. Artillery from Lavaca to Corpus Christi.[62]

David Bradbury was in charge of Singer's torpedo facility at Port Lavaca (later to take charge of all torpedo operations west of the Mississippi),[63] and Shea's reference to him seems to indicate that he and others involved in the local torpedo enterprise were then under Shea's jurisdiction. Because Shea singled out Bradbury and referred to him as "Captain," plans were probably in the works to take the Singer-Fretwell operation east. Singer and Fretwell were noticeably absent in this communication.

Shea's communication was the only document filed in William Longnecker's Confederate war record at the National Archives. No documents of any kind appear in either Edgar Singer's or John Fretwell's war record, and the only citation within these records having anything to do with their unique operation is a brief note stating that both men had been transferred to "The

Torpedo Bureau" in early 1863 (all other members of Shea's artillery had similar notations in their records, but only Longnecker's record contained a filed document).

To Confederate researchers this lack of original documentation is a source of constant frustration, for in all cases, regardless of subject, only bits and pieces seem to remain from original sources. For example, James Jones's invoice for $115 was discovered while searching the fragmented "Torpedo Papers" within the Confederate Navy Subject File at the National Archives, a branch of service to which Jones was not even attached. With the collapse of the Confederacy in 1865, literally tons of documents were intentionally burned or otherwise destroyed, and in many cases Confederate regimental muster rolls, order books, and personnel records were casually plucked from smoldering garbage heaps or found lying in gutters by advancing Federal troops looking for rebel souvenirs. Fortunately, in several cases these souvenirs were later donated to local historical societies in the decades following the war, and many eventually found their way to the National Archives.

From the evidence presented Singer and Fretwell probably had orders by early February 1863 to demonstrate some of their explosive devices to Gen. E. Kirby Smith's headquarters in Alexandria, Louisiana. By early February news of the sinking of the USS *Cairo* by a torpedo planted in the Yazoo River would have been well known west of the Mississippi, and officers within General Smith's command would certainly have wanted such devices for defending the Red River and other western waterways. Military commanders within the Trans-Mississippi Department were likely quite receptive to this novel form of warfare, so General Magruder probably would have dispatched Singer and Fretwell to General Smith's headquarters with their invention as soon as possible.[64]

Singer and Fretwell indeed reported to Trans-Mississippi Department headquarters in early February and were then requesting additional personnel to join them there: "Special Order 70. February 11, 1863. Messrs. Longnecker and James Jones members of Shea's Battalion have permission to report to Alexandria, Louisiana, and report to Lt. General E. Kirby Smith for the purpose of proposing the use of the Submarine Batteries of E. C. Singer and Company in Louisiana, and retain them in the employment of the government, or under such contracts as they may be able to effect for the use of the batteries."[65] Members of Singer's group slowly trickled eastward during the next few weeks.

From documentation available, it becomes clear that Singer and his men were constantly on the move, so it is difficult to pin down exactly when or where any certain individual or individuals were at any given time, although we do know that many members were east of the Mississippi River by mid-March.

Construction on the *Hunley* submarine must have commenced during the next several weeks, since it was completed in Mobile by early July 1863.[66] After meeting with General Smith, Singer and several of his associates were sent to southern Alabama "to report forthwith to General Maury, Commanding the District of the Gulf, with headquarters at Mobile."[67] Upon arriving in the blockaded port city of Mobile, Singer and Fretwell would have met with General Maury and outlined their plan to mine Mobile Bay with their newly developed torpedoes. Nothing is known of any demonstrations arranged for the general and his staff, because almost all Confederate records from Mobile were lost at the end of the war. But we do know that Mobile Bay was mined with Singer torpedoes by early summer.

With negotiations perhaps in progress Singer sent word to Texas for additional personnel: "Houston, March 17th, 1863. Special Order number 31. Private B. A. Whitney and J. D. Braman Company A, Shea's Battery Artillery are hereby detached for special duty for sixty days to assist in manufacturing and introducing Singer Sub-Marine Batteries and Torpedoes for the use of the Confederate States in such locations east of the Mississippi River as may be designated by the proper military authorities."[68]

Singer and Fretwell remained in Mobile for a relatively short period of time, for Whitney and Braman undertook contract negotiations with the military defenders of the city.[69] Since Singer and Fretwell had by then met with Generals Magruder, Smith, and Maury, they most likely would then have reported to government officials in Richmond. Singer and Fretwell traveled between Richmond and Mobile often during the next few months, and evidence exists regarding meetings conducted during March with government officials: "Under an act of the Confederate Congress approved on March 20, 1863, 'to provide and organize engineer troops,' Secretary of War James Seddon granted authority to Singer to form a company of twenty-five men or less, for special torpedo service, to be attached to the Bureau of Engineers under chief J. F. Gilmer, and subject to immediate orders of the commander of the district in which they should operate."[70]

B. J. Sage, one of the founders of the Confederate Secret Service, proposed other secret groups:[71]

> Such bands would work every variety of approved destructive expedient of the class held in view herein, known to warfare, or that may be invented and sanctioned—every variety of torpedo, and modes of working them defensively and offensively, submarine boats, and other contrivances of the kind. They could dash on vessels and steamboats in small boats, and capture them, or they might drench them with spirits of turpentine and other incendiary matter and burn them.
>
> Such a corps could move quickly from bay to bay, or river to river, and make frequent surprises . . . blow up railroad trains to prevent reinforcements

and supplies during a battle or an attack on a place held by the enemy. In short, they would effectuate their purposes by every variety of expedient and probable device and carry out many not now conceived of.... Such bands and the kind of warfare contemplated, must be authorized, so that persons engaged may have military status, and be protected by the government. Their captures should be secured to them as prizes. They should have a handsome bonus or bounty for the destruction of war vessels and transports (the bounty given now by act of April 19th, 1862, is only for destruction by novel means), and perhaps a small one for destroying any other crafts and property of the enemy. Respectfully, B. J. Sage.[72]

Sage was describing the future activities of Edgar Singer's group in remarkable detail, for the Texan and his organization carried out with great success all forms of covert activity mentioned, from planting underwater mines and railroad sabotage, to the construction and operation of submarine boats. H. N. Hill wrote about the official founding of Singer's unique organization:

Captain Singer then organized "Singer's Submarine Corps."; and among the men detached from military and naval departments for service with Singer were J. D. Braman, R. W. Dunn, H. L. Hunley, B. Gus Whitney, D. Bradbury, Jimmy Jones, C. E. Frary, J. R. Fretwell, F. M. Tucker, L. C. Hirshberger and others—the majority serving throughout the war. The company was furnished by the authorities with the necessary ammunition and materials for manufacturing the devices, as well as free transportation for both men and machines. Their compensation was 50 percent of the value of all vessels of war and other federal property destroyed by means of their inventions, and all arms captured by them; special rewards were also provided by Congress.[73]

Of the men named, Horace Lawson Hunley did not become a member of the organization until sometime in March, when the group moved the hub of their operations to Mobile.[74] Some ten months earlier, following the Federals' capture of New Orleans in late April, H. L. Hunley, James R. McClintock, Baxter Watson, and Henry Leovy arrived in Mobile with hopes of continuing their submarine experiments. The four partners had designed and built the small three-man privateering submarine *Pioneer* some weeks before the loss of New Orleans and had been forced to scuttle their underwater invention and retreat to Mobile to continue their venture. With a privateering commission from the Confederate government in hand, they approached the military authorities of Mobile, presented their unusual plans, and were immediately granted facilities at the Park and Lyons machine shop near the harbor on Water Street (the same facility where Singer and Fretwell would manufacture their new underwater contact mine).

Some weeks after their arrival in Mobile, McClintock and Watson were directing new designs for a second submarine boat, radically different from the one they were forced to scuttle in New Orleans. At the same time, Leovy and Hunley (along with an unknown gentleman named Duncan) journeyed to Richmond to negotiate a blockade-running contract with the War Department.[75] Leovy and Hunley planned (and perhaps set up) a blockade-running venture in October 1862, according to a War Department document stating that "Leovy, Hunley and Duncan" were under contract with "the Secretary of War" to deliver "Foreign Supplies to the Ordnance Bureau."[76] Nothing more has been documented of the ultimate success or failure of the Leovy-Hunley venture.

Hunley financed the submarine's construction at the Park and Lyons machine shop from his personal accounts.[77] Among the many soldiers on detached duty at the facility was Lt. William Alexander, a young mechanical engineer who had migrated to Alabama from London just two years before the Civil War began. At the commencement of hostilities, Alexander enlisted in Company A of the 21st Regiment Alabama Infantry, a unit recently assigned to garrison duty at Fort Morgan guarding the mouth of Mobile Bay.[78] Within weeks after their deployment, Alexander and several other skilled machinists and engineers were detached from their regiment and placed on temporary duty at the Park and Lyons machine shop. For several weeks Alexander and his machinists had been boring out the barrels of old prewar Mississippi sporting rifles to make them compatible with military ammunition.[79] With the unexpected arrival of Watson, Hunley, Leovy, and McClintock, Lieutenant Alexander's superiors ordered him to place the conversions on hold and give his full attention to the unique project proposed by the inventors.

Apparently the Louisiana partners attempted to build an underwater craft far too advanced for the technological times, as noted in a letter written by James: "We built a second boat at Mobile, and to obtain room for machinery and persons, she was made 36 feet long, Three feet wide and four feet high. Twelve feet of each end was built tapering or molded, to make her easy to pass through the water. . . . There was much time and money lost in efforts to build an electro-magnetic engine for propelling the boat."[80] The plan to power their recently designed submarine with an electric motor could be viewed as unimaginable considering the era in which they were working. Early electric motors capable of propelling such a vessel had existed since the Victorian Age, and Singer agent Baxter Watson sent a proposal to the Confederate War Department stating that an engine of "electro-magnetism" capable of powering such a craft could be purchased in New York or Washington, D.C., at a cost of five thousand dollars.[81]

McClintock stated in his letter that the electric motor designed and fabricated for propelling their second submarine boat "was unable to get sufficient power to be useful." Unfortunately, McClintock remained silent about how the group tested their electric motor, so we do not know how close they came to fabricating the world's first electrically powered submarine. Surviving documentation suggests that they experimented for several months designing and redesigning the power source by which the submarine was to have been propelled.[82]

When the electric motor failed, the determined inventors turned to a more practical means of propulsion, a small custom-built steam engine.[83] No documentation explains how smoke may have been discharged, but the steam engine may have been designed to build up great pressure within its boiler while running on the surface. Prior to submerging, the crew extinguished the fire, causing the submarine to be propelled with the remaining pressure stored in the boiler. Both McClintock and Watson had been steam gauge manufacturers so would have known the system's limitations. Obviously they must have carefully weighed its feasibility to have attempted its use. In the years following the Civil War this same steam system was incorporated in an experimental European submarine with great success.[84]

However, steam propulsion also proved inadequate, conceivably due to the scarcity of suitable steam components in wartime Mobile. The designers removed the failed steam engine and installed a propeller shaft designed to be turned by four men. By mid-January 1863 (at about the same time that Singer and Fretwell were planning to move their torpedo operations eastward) the *American Diver* (as a Confederate deserter called the submarine[85]) was ready for sea trials in Mobile Bay. McClintock himself later wrote some years after the war, "I afterwards fitted cranks to turn the propeller by hand, working four men at a time, but the air being so closed, and the work so hard, that we were unable to get a speed sufficient to make the boat of service against vessels blockading this port."[86] Lt. William Alexander tells of the submarine's fate: "It was towed off Fort Morgan, intended to man it there and attack the blockading fleet outside, but the weather was rough, and with a heavy sea the boat became unmanageable and finally sank, but no lives were lost."[87]

Fortunately for the disappointed, out-of-work submarine designers, torpedo experts Edgar Singer and John Fretwell arrived in Mobile within weeks after the loss of the *American Diver* and undoubtedly were keen on seeking out competent local engineers and machinists familiar with Mobile Bay to join them in their proposed torpedo operations. Since Hunley, Watson, Leovy, and McClintock had previously experimented with underwater explosives for use with their submarine vessels and were accomplished inventors, the four men may well have been regarded as a godsend to the newly arrived Texans armed only with drawings and diagrams of their new torpedo system. For nearly a year the four partners had interacted with Mobile military authorities, undoubtedly gaining invaluable experience about how to cut through local bureaucratic red tape to obtain scarce manufacturing materials needed for their operation.

Singer and Fretwell soon approached McClintock, Watson, and Hunley and offered membership in Singer's torpedo organization. Henry Leovy's name was absent from known group rosters.[88] Leovy was absent because "he went to Richmond and was appointed by the president, commissioner to examine and settle disputes between the civil and army authorities in southwestern Virginia. In many parts of western Virginia and North Carolina spies and deserters were creating alarm, and many arrests were necessarily made. The duty was an arduous one, and for his successful services he was promoted and commissioned to the rank of colonel of cavalry, and later was assigned to duty as judge of the military court for the southwestern Virginia district."[89] Although Leovy's duties called him away from Mobile, he never lost contact with his old partners and apparently could be considered a fringe member of the Singer group since he became a financial partner in one of the organization's most daring projects.[90] With his wife comfortably resituated in Abbeville, South Carolina (a western region of the state far removed from the war), Leovy continued with government duties throughout the Confederacy.

With the incorporation of new members Hunley, McClintock, and Watson the small company of underwater explosive experts quickly turned their attention to contract negotiations with Mobile military authorities and expansion of their operation to regions throughout the South. While Mobile Bay was being surveyed and torpedo materials gathered, discussions concerned fabrication of yet a third vessel. The three new Singer members may well have been regarded as the leading submarine experts then alive. Five men from the group purchased shares in the proposed project. Singer purchased one-third of the vessel at a cost of five thousand dollars; Hunley retained another third, with the remaining shares being evenly divided between Dunn, Whitney, and Braman, bringing the total cost of the submarine boat to a staggering fifteen thousand dollars, quite a considerable sum of money in 1863.[91]

With financial backing in place, construction began immediately at the Park and Lyons machine shop under the direction of James McClintock. Engineering officer Lieutenant Alexander provides a firsthand description of this innovative diving machine soon to be christened the *H. L. Hunley* after the organization's leading submarine advocate and financier.

> We decided to build another boat, and for this purpose took a cylinder boiler which we had on hand, 48 inches in diameter and twenty-five feet long (all dimensions are from memory). We cut this boiler in two, longitudinally, and inserted two 12-inch boiler iron strips in her sides; lengthened her by one

tapering course fore and aft, to which were attached bow and stern castings, making the boat about 30 feet long, 4 feet wide and 5 feet deep. A longitudinal strip 12 inches wide was riveted the full length on top. At each end a bulkhead was riveted across to form water-ballast tanks (unfortunately these were left open on top); they were used in raising and sinking the boat. In addition to these water tanks the boat was ballasted by flat castings, made to fit the outside bottom of the shell and fastened thereto by "Tee" headed bolts passing through stuffing boxes inside the boat, the inside end of the bolt squared to fit a wrench, that the bolts might be turned and the ballast dropped, should the necessity arise.

In connection with each of the water tanks, there was a sea-cock open to the sea to supply the tank for sinking; also a force pump to eject the water from the tanks into the sea for raising the boat to the surface. There was also a bilge connection to the pump. A mercury gauge, open to the sea, was attached to the shell near the forward tank, to indicate the depth of the boat below the surface. A one and a quarter inch shaft passed through stuffing boxes on each side of the boat, just forward of the end of the propeller shaft. On each side of this shaft, outside of the boat, castings, or lateral fins, five feet long and eight inches wide, were secured. This shaft was operated by a lever amidships, and by raising or lowering the ends of these fins, operated as the fins of a fish, changing the depth of the boat below the surface at will, without disturbing the water level in the ballast tanks.

The rudder was operated by a wheel, and levers connected to rods passing through stuffing-boxes in the stern castings, and operated by the captain or pilot forward. An adjusted compass was placed in front of the forward tank. The boat was operated by manual power, with an ordinary propeller. On the propelling shaft there were formed eight cranks at different angles; the shaft was supported by brackets on the starboard side, the men sitting on the port side turning on the cranks. The propeller shaft and cranks took up so much room that it was very difficult to pass fore and aft, and when the men were in their places this was next to impossible.

In operation, one half of the crew had to pass through the fore hatch; the other through the after hatchway. The propeller revolved in a wrought iron ring or band, to guard against a line being thrown in to foul it. There were two hatchways—one fore and one aft—16 inches by 12, with a combing 8 inches high. These hatches had hinged covers with rubber gaskets, and were bolted from the inside. In the sides and ends of these combings glasses were inserted to sight from. There was an opening made in the top of the boat for an air box, a casting with a closed top 12 by 18 by 4 inches, made to carry a hollow shaft. This shaft passed through stuffing boxes. On each end was an elbow with a 4 foot length of 1–½ inch pipe, and keyed to the hollow shaft; on the inside was a lever with a stop-cock to admit air.[92]

Alexander's description of the *Hunley* was based on forty-year-old memories, and he apparently had forgotten several features, such as the vessel's true length. Although Singer himself, when later interviewed, had forgotten some details, both men's recollections, when taken together, paint a much clearer picture of the vessel and its unique characteristics:

> Captain Singer says this boat was the invention of H. L. Hunley and was built under the supervision of James McClintock. It was owned jointly by Hunley, Braman, Whitney, Singer and Dunn, and was valued at $15,000.... Captain Singer endorses the following description of the *Hunley*: She was built of old boiler iron impervious to water or air. She was forty feet long, with a beam of five or six feet, about six feet depth of hold, and resembling a cigar, tapering at both ends and riveted.
>
> She was propelled by a shaft, or screw, which ran horizontally along her hold almost stem to stern and was turned by the manual force of eight men.... The commander manipulated the torpedo. A circular hatchway two feet in diameter, with a low combing around it, was placed well forward and covered by an iron cap, hinged and airtight. A glass bull's-eye inserted in the cap supplied the periscope. Water tight compartments, filled or emptied at will, enabled the boat to rise or submerge....
>
> Besides an ordinary rudder, the vessel was equip[ped] with side paddles, or fins, which served to guide it up and down with reference to the surface of the water. In preparing for action a floating torpedo was secured to her stern by a line over a hundred feet long, the crew embarked, the hatchway was closed, the hawsers were ordered off and the men began revolving the shaft.
>
> The Hunley's greatest speed did not exceed four knots. She could remain submerged for an hour or so without any great inconvenience to her crew, although no means were provided for procuring fresh air after the hatch was closed. The plan of attack was to dive beneath an enemy's ship, hauling the torpedo after her; the bomb's triggers, or sensitive primers, would thus press against the ship's bottom, exploding the charge and inevitably sinking the ship. The Hunley used the Singer torpedoes exclusively.[93]

Sometime during the construction of the *Hunley* or the *American Diver* Lt. George E. Dixon, an officer in the 21st Alabama, enters the story of the *Hunley*. A Kentucky native and mechanical engineer, Dixon had entered Confederate service in the spring of 1861, leaving his position as a Mobile steamboat engineer to enlist in the 21st Alabama Regiment. Due to a severe leg wound received at the Battle of Shiloh, he may have walked with a limp and was thus relieved of active field service. He also was likely on detached duty with Alexander at the machine shop during the construction of both the *American Diver* and *Hunley*. His company commander mentions in a letter that military authorities had assigned Lieutenant Dixon to the *Hunley* project.[94]

Formation and Deployment

With monies in place and McClintock apparently in charge of overseeing the submarine's construction, Singer, Fretwell, and the other Texans discussed the future activities of their organization. Several operatives under the direction of Fretwell were quickly sent to Mississippi to help Capt. Isaac Brown (hero of the CSS *Arkansas*) defend the water approaches leading to the Confederate shipyards at Yazoo City:

> Jackson, Miss. April 1, 1863. Messrs John R. Fretwell & Co. propose to plant two of Singer's Submarine Batteries in the Tallahatchie at a point to be designated by the Commander of the District. They wish a requisition upon the Ordnance Department for the powder to charge the Batteries, also upon the Quartermasters for Transportation for themselves and materials to and from the point where they are to be planted. If they succeed in blowing up or destroying one or more of the enemy's gunboats they are to be paid for each gunboat Twenty thousand dollars, for each transport with Troops on Board Ten thousand Dollars, for each transport without troops on Board Five Thousand Dollars.[95]

Obviously, intense negotiations between Singer operatives and government officials had taken place, since even the reward for destroying empty troop transports had been both debated and agreed upon. Fretwell and his associates labored in Mississippi at least the next four months. According to postwar writings of Captain Brown, Fretwell's arrival could not have come at a better time. Federal warships were poised to commence operations against the Confederate shipyards at Yazoo City, so Brown needed a reliable torpedo (better than the crudely constructed glass demijohn affair used to sink the USS *Cairo* some four months earlier) to obstruct the Yazoo River.

Although the order states that the Tallahatchie River was the intended waterway to be obstructed by Fretwell's group, the proposed operation was quickly expanded to include the Yazoo as well. Captain Brown, then assigned the task of defending all waterways within the state of Mississippi, commented about the manufacture and deployment of Singer mines: "I also made a contract with Dr. Fretwell and Mr. Norman, then at Yazoo City, for fifty or more of these destructives [torpedoes] on Dr. Fretwell's plan—automatic action on being brought in contact with a vessel or boat."[96]

Captain Brown was apparently quite impressed with Fretwell's new underwater explosive device, for fifty torpedoes at the agreed-upon price of $150 apiece would cost the Confederate government $7,500. In Brown's mind this was obviously a fair price to pay to protect the valuable Yazoo shipyard.[97]

Although we can only speculate about who accompanied Fretwell on his torpedo mission to Mississippi, Hunley was among them.[98] Proof establishing Horace Hunley's presence in Mississippi at this time comes from a contemporary Hunley letter written in the rural town of Canton, Mississippi, to Henry Leovy, discussing the current situation at Vicksburg.[99]

While Fretwell and associates were surveying Mississippi waterways and preparing facilities in which to fabricate their deadly devices, in Mobile Singer agents Whitney and Braman were nearing the end of contract negotiations with military authorities to mine Mobile Bay. They signed the following contract on April 17, just two weeks after Fretwell had departed for Mississippi:

> Headquarters District of the Gulf. Mobile, Ala. April 17, 63. Mr. B. A. Whitney & J. D. Braman agents for E. C. Singer & Co. of Lavaca, Texas will proceed to make and plant under the directions of the military commanders of the district one Hundred of their submarine batteries. Upon locating and planting these under orders or upon delivery to the Ordnance Department, when the commander desires delay, they will be paid for by the Ordnance Officer at the rate of One hundred & fifty dollars for batteries in numbers not less than five in one payment.
>
> Requisitions upon the Ordnance Department for tin, powder, caps, and anchors for making and planting will be honored without charge to the company as also all transportation of the battery's material and the parties planting them will be furnished by the Quartermasters under orders of the Commander S. B. Buckner, Major General Commanding.[100]

During these contract negotiations, Fretwell was then in Mississippi, perhaps with Hunley and several other Singer operatives. A letter written by Robert W. Dunn to General Magruder places Singer himself in Richmond, meeting with officers of the Engineering Department: "Since your permission to report to General E. Kirby Smith, at Shreveport, Louisiana . . . we were permitted by him to report to the Secretary of War at Richmond, Virginia. . . . Upon presenting certificates of the engineers of your department, as also others from General Buckner and Ledbetter, of tests made in all the departments named, it was deemed by the Secretary of War to give us every possible facility for using our torpedoes. We were at once transferred by him to the Engineer Troops."[101]

Singer and those associated with his torpedo operations were assigned to the engineers to establish military credentials for group members in case of capture.[102] Although the years 1861–65 are thoroughly documented in the order books, the number of communications sent to "E. C. Singer and Company" are surprisingly few. Most orders directed to the Singer group are rather ordinary and uninteresting and shed little or no light on group member whereabouts, current operations, funding, transportation expenses, or any of the various meetings known to have taken place in Richmond. Although it is well documented that Singer and several of his agents traveled back and forth from Richmond often, not one order even hints that any of his men were ever summoned to meetings in the capital.[103]

It seems highly likely that Singer and his organization received their orders from a government agency as yet unknown and undocumented. The wartime writings of B. J. Sage provide some information:[104]

> It was finally proposed . . . to organize them [Bands of Destructionists and Captors] under the Engineer Bureau, according to an act approved on March 19th, 1863. This was adopted, and the men were individually authorized to proceed to the department or district where they wished to operate, and be enlisted and organized with an engineer company; and then be detailed for their special purpose—the commanding General being requested or recommended to furnish them with transportation, work shop aid, ordnance stores, and military protection and assistance, at his discretion.
>
> Under this plan, some of the most ingenious, enterprising, daring and patriotic men in the Confederacy have gone to the south and southwest, who will form the nuclei of many organizations, of the progress of which, in the work of destruction, I hope we shall soon hear favorable and loud reports.[105]

"Bands of Destructionists" like the one headed by Singer and Fretwell had a unique relationship with established military authorities, for commanders of districts were only "requested or recommended" to give support. Sage goes on to state that "the nuclei of many organizations" had thus far been established throughout the South, suggesting that some secret umbrella organization must have been in place to oversee operations.

Evidence that the mysterious B. J. Sage was the man in charge of the organization given the task of directing the activities of groups such as Singer's is found in a captured document signed by Sage himself. The secret communication was a letter of introduction lost by a Singer operative while crossing the Mississippi River and identifies all Confederate agents, saboteurs, and informers then working behind enemy lines in that region of the South.[106] Within weeks after this list of rebel operatives was captured, Federal authorities began rounding up the men named; some modern historians call the capture of this document the biggest blunder ever to befall the Confederate Secret Service.[107]

To complicate matters, military ranks assigned to those associated with Singer and other such organizations are somewhat confusing. Edgar C. Singer, Horace L. Hunley, Robert W. Dunn, and David Bradbury at one time or another during the war years were all referred to in official orders, reports, and surviving documents as "Captain" (Hunley himself actually signed several surviving military requisitions as "Captain H. L. Hunley").[108] However, none of these individuals appear on existing Confederate rolls as having ever received a commission. A late war document from the Confederate Adjutant and Inspector General's Office discusses the secret assignment of a John P. Halligan, whose name is also absent from Confederate commission rolls: "Special Order Number 259. Second Lieutenant J. P. Halligan, of a secret service company, will report for temporary duty with flag-officer E. Farrand, C.S. Navy, at Mobile Alabama."[109] There is no current evidence linking Halligan to the Singer organization.

It thus appears that members of secret service companies were assigned ranks just as individuals in other branches of military service were. Singer himself stated, "The appointment of additional corps, commission and compensation being similar to those granted the Singer band" occurred throughout the South.[110] He also noted that inventors such as "T. E. Courtenay, Colonel Hill, Major Perkins, Hunter, Weldon and others" commanded these units. T. E. Courtenay (rank not cited) commanded Courtenay's Secret Service Corps and is referred to in surviving Confederate documentation as "Captain Courtenay."[111] Zedekiah McDaniel and his assistant Francis Ewing had sunk the USS *Cairo* in the Yazoo River with a crude mine in December 1862. McDaniel commanded a twenty-eight-man secret service company known as McDaniel's Secret Service Corps, and he was also officially referred to in surviving documentation as Captain McDaniel.[112]

Payments to individuals associated with covert or clandestine units were established and funded from a congressional "act to pay officers, non-commissioned officers and privates not legally mustered into the service of the Confederate States, for services actually performed." William Tidwell, perhaps the leading expert on the Confederate Secret Service, cites this act as proof that individuals associated with secret service companies were in fact assigned official rank but were never "legally mustered into the service of the Confederate States."[113] It seems reasonable to assume that Edgar Singer and several individuals within his command received some form of official rank from the Confederate government, but the surrounding details remain undocumented and vague.[114]

The name "Singer Secret Service Corps" does not appear on any known documents until several months later, so when the group received this designation is not known.[115] We know that the similar twenty-member group Courtenay Secret Service Corps was in existence by August 1863,[116] and Captain Courtenay himself designed various explosive devices similar to those utilized by Singer's organization.

While construction on the *Hunley* was progressing favorably at the Park and Lyons machine shop, David Bradbury and his assistants were diligently building torpedoes in Port Lavaca for the defense of the Texas coast. The success of the Singer torpedo had apparently made a great impression on Singer's old commanding officer, for in May 1863, Major Shea ordered eighteen additional mines to be anchored near Fort Esperanza.[117]

From a report written by Capt. David Bradbury, we get some idea of problems associated with anchoring Singer torpedoes in deep water and the deterrent factor of deploying such devices in the defense of the Texas coast: "By order of Major Shea, I placed 18 floating torpedoes in the channel between Fort Esperanza and the bar at Pass Cavallo, but the channel was very wide, water deep (30 feet) and current strong, and I made up my mind when I put them there that it was doubtful if they ever did any good. . . . I have good reason to believe that the fact of their being there, and they knowing it [the enemy], has kept them out."[118]

One of the problems associated with the Singer-Fretwell mine in salt water was that its exposed spring-loaded detonating system became useless within several weeks after deployment. After lying dormant beneath the surface of the water for four weeks or more, the cocked detonating rod would be covered in barnacles and tiny shells, which prevented the mechanism from snapping against the end of the container and detonating the powder.[119] The spring itself, even when the torpedo was deployed in fresh water, lost much of its tension after being cocked and submerged for extended periods of time. Surely, the system's inadequacies were a closely guarded secret, kept from all but the highest-ranking officials. Captain Bradbury reported that "the fact of their being there, and they knowing it [the enemy], has kept them out" of Matagorda Bay.[120]

While Captain Bradbury and his men diligently worked to manufacture Singer torpedoes for use west of the Mississippi River, Singer and his associates at Richmond were carefully drawing up detailed diagrams of their unique explosive device for the Confederate Patent Office. On May 18, 1863, Singer received a disappointing response from the Confederate Commissioner of Patents Rufus Rhodes that stated, "Your application for a patent for a torpedo has been examined. . . . All you can claim as your invention are the mechanical elements."[121]

Despite Singer's apparent patent frustrations, he lost little time in deploying several of his torpedo mechanics and operatives south to the port city of Wilmington, North Carolina. A government invoice drawn at Wilmington on May 13, 1863, states that fifteen hundred dollars was paid for ten Singer torpedoes.[122] Dunn's letter to General Magruder states, "We were at once transferred by him [the Secretary of War] to the Engineer Troops and ordered to report to General Joseph E. Johnston at Morton, Mississippi. Leaving operators at Richmond, Wilmington, Savannah, Mobile, and Charleston."[123] The Singer operatives were probably sent to Wilmington, Savannah, and Charleston soon after the meetings in Richmond. There is a question of torpedo operations having taken place in Savannah; no surviving documentation other than the Dunn letter even hints that operations were ever conducted in that Georgia coastal city at anytime during the war.

That Singer agents apparently set up a previously unknown operation in Savannah should not be surprising because organizational stealth and secrecy were essential to the group's success, so many of the daring covert operations, adventures, and schemes hatched by the Singer Secret Service Corps will remain a mystery.

In some cases historians have obtained information about the very existence of new destructive inventions, secret congressional meetings, and railroad sabotage behind enemy lines from such widely ranging sources as a sentence from a letter or scribbled summary appearing at the bottom of a military requisition for supplies. A "complete and comprehensive" history, covering all operations and covert missions attempted by the various members of the Singer organization, ever being compiled from the fragmented sources now available is highly unlikely.

Adding to this confusion is an extremely valuable, confidential communication penned in early 1864, in which Dunn outlined some of the accomplishments already achieved by Captain Singer and his operatives. While briefly touching on early operations conducted in Louisiana, Dunn made the following claims regarding the first enemy vessel apparently sunk by a Singer-Fretwell torpedo: "Our success in the use of the stationary torpedo, has been the destruction of the enemies transport Gray Cloud in the Atchafalger Bay, Louisiana, in June [1863] last, killing and drowning 140 men."[124]

However, surviving Union naval records state that the *Gray Cloud* sank at the mouth of the Atchafalaya River after hitting a submerged "snag" and that the incident happened with no loss of life. On March 14, 1863, the Federal steam sloop *Richmond* received minor damage after striking a submerged mine (of unknown origin) at Port Hudson, Louisiana, but according to surviving documentation, the USS *Richmond* remained the only Union vessel to have struck a submerged enemy torpedo in Louisiana waters until the spring of 1864.[125]

Although no corroborating documentation has thus far come to light regarding the sinking of a Union troop transport in Atchafalaya Bay, the alleged event must have been based on a partial Singer success since it is highly doubtful that Dunn would purposely fabricate such an important event for the military commander of Texas, New Mexico, and Arizona. Apparently Dunn considered the *Gray Cloud* the first enemy vessel sunk by a Singer-Fretwell torpedo. Although Dunn was in Shreveport in June 1863,[126] nothing more is known about the sinking of this Union vessel. Even though the Dunn letter appears to have some minor contradictions, the vast majority of the confidential communication deals with facts and events that can be confirmed from surviving historical records, and we now have a much richer accounting of the secret activities conducted by Singer's agents, operatives, and engineers.

While Fretwell and his associates were hastily setting up a torpedo manufacturing facility in Yazoo City, and McClintock and staff were poring over submarine diagrams in Mobile, Singer agents Whitney and Braman were nearing completion of the first twenty torpedoes designated for use in Mobile Bay. Surviving documentation suggests that Singer-Fretwell torpedoes and the *Hunley* were built side by side by the same Singer personnel and that several would soon find themselves making up the bulk of the little submarine's untested crew.

2 First Torpedo Strike and Launching the *Hunley*

THE LATE SPRING AND EARLY SUMMER of 1863 was a chaotic time for both the struggling Confederacy and the newly inducted, and untested, members of the Singer Secret Service Corps. While Edgar Singer's operatives were busily fabricating their submarine *H. L. Hunley* and setting up torpedo manufacturing facilities in Mobile and Yazoo City, the Confederate armies in both the eastern and western theaters were about to experience dreadful defeats from which the isolated South would not recover. While Singer and several of his associates were attending meetings in Richmond with representatives of the Engineering Department and Gen. Gabriel Rains's recently established Torpedo Bureau, operatives Whitney and Braman were preparing to deploy several Singer torpedoes in Mobile Bay.[1]

On July 1, 1863 (at exactly the same time that Gen. Robert E. Lee was hastily funneling troops onto the field for the first day of battle at Gettysburg, Pennsylvania), Confederate engineering general Danville Leadbetter was writing a dispatch at his headquarters in Mobile: "Engineer Office, Mobile, July 1, 1863. To Captain H. Myers, Ordnance Officer. Captain: Messrs Whitney and Braman have manufactured and planted in the waters of Mobile Bay, twenty torpedoes, for which they are to be paid, as directed by General Buckner, one hundred and fifty dollars each. Respectfully D. Leadbetter, Brig. General Engineers."[2]

Thus, Singer's Mobile-based torpedo operations were proceeding as planned, and the military authorities responsible for overseeing activities were pleased with the end product. Just how many Singer operatives were then working on torpedoes in the sweltering heat of the Park and Lyons machine shop is unknown, but judging from the great quantity of mines fabricated during the weeks that followed, we can assume that several of Singer's recently transplanted Texans had been assigned to the Mobile project.

While Whitney and Braman oversaw the fabrication of Singer torpedoes, work on the *Hunley* submarine was progressing favorably under the watchful eyes of James McClintock and Lt. George E. Dixon and Lt. William Alexander.[3] By late June the hull of the submarine would have been nearing completion, which perhaps caused some of the Texans assigned to the project to face the sobering reality that they would soon be needed to crew the sinister-looking vessel and put it through underwater harbor trials.

As work on the submarine was nearing completion, John Fretwell and his associates in Mississippi were kept constantly informed of the submarine's status. Fretwell continually shared this confidential information with Capt. Isaac Brown, who then brought the submarine to the attention of General Beauregard, military commander of Charleston.

For several months prior to setting up the torpedo manufacturing facility on the Yazoo River, General Grant and an army of nearly seventy thousand men had ravaged the region and laid siege to nearby Vicksburg, the last southern stronghold on the Mississippi River. By means of a Union trench line some fifteen miles long, Grant had succeeded in sealing off the city, causing one Confederate soldier to write later, "A cat could not have crept out of Vicksburg without being discovered."[4] While the Union Mississippi Squadron held the river, Grant's mortar batteries lobbed exploding shells into the city, forcing the civilian population to live in cellars and nearby caves. As the weeks dragged on, food supplies in the isolated city grew increasingly short, which forced many of the garrison troops and desperate civilians trapped within to eat mules, rats, and stray dogs that came within their reach.[5]

News of the desperate conditions apparently had a sobering effect on the recently inducted Singer operative Horace Hunley, for soon after Grant's forces had bottled up the Vicksburg garrison behind their fortifications, he sent the following letter to his old friend Henry Leovy (who may have been in Richmond at the time):

> Canton Mississippi, May 4th, 1863. My Dear friend Henry: I have some idea of joining our forces in defense of Vicksburg, and should any accident befall me, I wish you to take charge of my affairs and manage them and after disposing of the sugar and tobacco use the money for speculation or running the blockade till the end of the war.... I gave A. J. Picton two thousand dollars for the payment of freight charges.... Dixon [undoubtedly Lt. George E. Dixon] was indebted to me about $82.00 on the old score before his last trip on the Mobile and Ohio railroad. Your Friend. H. L. Hunley.[6]

The letter confirms that Hunley was in Mississippi at the same time that Fretwell and his associates were and proves to be an excellent example of Hunley's overzealous dedication to duty. Hunley probably never attempted to cross General Grant's encircling trenches around Vicksburg, but the letter indicates that he was seriously contemplating an attempt to enter and join the garrison troops in Vicksburg. The fact that Hunley requested Leovy "to take charge of my affairs and manage them" in case of his death confirms that Hunley was seriously considering a course of action that could be considered above and beyond the call of duty.

On July 4, 1863, Gen. John Pemberton surrendered the city of Vicksburg to Ulysses S. Grant. Five days later the last southern stronghold on the Mississippi River, Port Hudson, Louisiana, fell to Union forces, thus bringing the entire length of the river under Federal control. From then until the end of the war, the Confederacy was effectively split in two, with armies east and west of the Mississippi River operating independently from one another. Although there would be two more years of brutal war, many in the South viewed the double defeats in July 1863 (Gettysburg and the loss of Vicksburg and the Mississippi) as the beginning of the end for the ill-equipped and isolated Confederacy.

Within a week after accepting the surrender of Vicksburg and its nearly thirty thousand half-starved defenders, General Grant turned a wary eye toward the Confederate shipyards at Yazoo City and sent a directive to Admiral Porter to send some gunboats up the Yazoo River: "Vicksburg, Miss., July 11, 1863. Admiral Porter, Commanding Mississippi Squadron. Sir: General Washburn informs me that the Yazoo River has risen 6 feet. Will it not be well to send up a fleet of gunboats and some troops and nip in the bud any attempt to concentrate a force there? I will order troops at once to go aboard the transports. Very Respectfully, U. S. Grant."[7]

With an affirmative response from Admiral Porter, General Grant relayed the following message to General Herron: "Vicksburg, Miss., July 11, 1863. Major-General F. J. Herron. Sir: You will proceed with your command on transports to Yazoo City, take possession of that place, and drive the enemy from that place and section. Johnston is reported to have sent orders to have Yazoo City fortified. This we can not permit. Admiral Porter is sending gunboats to cooperate. Communicate with him, and move when he is in readiness. Take with you a battery, if you can get it aboard without too much delay. By order of Major-General U. S. Grant."[8]

As General Grant and Admiral Porter prepared to move on Yazoo City, with its valuable stores of cotton and munitions, Gen. Joseph E. Johnston was sending the following orders to Capt. Isaac Brown: "Headquarters, Jackson, Miss., July 14, 1863. Captain I. N. Brown. Sir: If it is necessary to abandon the Yazoo country, you will destroy all steamboats and public property to the extent of your means. By Command of General Johnston."[9] With a Union squadron of gunboats and troop transports heading their way, Captain Brown wasted little time in directing Fretwell and his staff to quickly mine the approaches to the city with as many torpedoes as were then available at their manufacturing facility.

Years later Brown himself wrote about his activities: "On the morning of the Union advance upon Yazoo City [July 13, 1863], I had myself placed two of these 'Fretwells' half a mile below our land batteries."[10] It seems that he had great faith in the destructive power of a Singer-Fretwell torpedo, for he stated quite literally that he had gone out of his way to deploy two of them himself.

Under the direction of Captain Brown and his officers, the hectic dismantling of the Yazoo City Navy Yard and the vessels attached to it began immediately. Valuable naval equipment and all the munitions that could be assembled were hastily piled on wagons for transport to Selma, Alabama, as ragged soldiers prepared to torch then useless troop transports at their moorings. While Confederate sailors prepared to set fire to anything that could not be transported eastward, Fretwell's men were busily priming their recently completed torpedoes with gunpowder and anchoring them in the muddy river some miles south of Yazoo City. The time to prove the military worth of their deadly underwater invention had at last arrived.

Adm. David Porter sent Union secretary of the navy Gideon Wells a report about the success of the joint army and navy operation into rebel territory up the Yazoo River:

> Flagship Black Hawk, off Vicksburg, July 14, 1863. Sir: Hearing that General Johnston was fortifying Yazoo City with heavy guns, and gathering troops there for the purpose of obtaining supplies for his army from the Yazoo country; also, that the remainder of the enemy's best transports were there, showing a possibility of his attempt to escape, Major-General Grant and myself determined to send a naval and military expedition up there to capture them. The Baron De Kalb, New National, Kenwood and Signal were dispatched under command of Lieutenant-Commander John G. Walker, with a force of troops numbering 5000 under Major-General Francis J. Herron.
>
> Pushing up to the city, the Baron de Kalb engaged the batteries, which were all prepared to receive her, and, after finding out their strength, dropped back to notify General Herron, who immediately landed his men, and the army and navy made a combined attack on the enemy's works; the rebels soon fled, leaving everything in our possession, and set fire to four of their finest steamers that ran on the Mississippi River in times past.
>
> The army pursued the enemy and captured their rear guard of 260 men, and at last accounts were taking more prisoners. Six heavy guns and one vessel, formerly a gunboat, fell into our hands, and all the munitions of war.[11]

These confident statements regarding capturing troops and burning steamers were soon tempered with sobering news of the loss of a powerful gunboat sent to the bottom of the Yazoo River by a Singer torpedo, an infernal underwater explosive device that the admiral himself harshly condemned as "some new invention of the enemy."[12] Admiral Porter's report continues:

> Unfortunately, while the Baron de Kalb was moving slowly along, she ran foul of a torpedo, which exploded and sank her. There was no sign of anything of the kind to be seen. While she was going down another exploded under her stern. The water is rising fast in the Yazoo, and we can do nothing more than

get the guns out of her and then get her into deep water, where she will be undisturbed until we are able to raise her.

But for the blowing up of the Baron De Kalb, it would have been a good move. We have generally obtained information of torpedoes from Negroes and deserters, but heard nothing of this. Many of the crew were bruised by the concussion, which was severe, but no lives were lost. The officers and men lost everything. She went down in fifteen minutes. We must have her up again as soon as possible. We have much to contend with these narrow rivers, and can not guard against these hidden dangers while an enemy's flag floats. The usual lookout was kept for torpedoes, but this is some new invention of the enemy, which we will guard against hereafter. Your Obedient Servant, David D. Porter, Acting Rear-Admiral, Commanding Mississippi Squadron.[13]

As the Union army and navy approached, Fretwell's torpedo engineers had little time to celebrate their first victory and most likely hastened to join Captain Brown and his officers in their eastward retreat to Selma, Alabama. Seven days later Brown's force regrouped in Meridian, Mississippi, where General Johnston received news of the sinking of the *Baron DeKalb* and loss of Yazoo City: "Meridian, July 21, 1863. General Johnston: We repulsed the gunboats at Yazoo City on Monday last, but our infantry force retreated and I had to abandon my guns. We destroyed our steamboats. The ironclad De Kalb, of 13 guns was sunk by torpedo. I will report in writing from Selma. My few men have gone to Mobile. Isaac N. Brown, Commander, C.S. Navy."[14]

Upon arrival in Mobile, Fretwell and associates would have joyously shared the news of their sinking of a Federal gunboat with old friends Gus Whitney and J. D. Braman, who were themselves busily overseeing work on Singer-Fretwell torpedoes. Within hours after the arrival of the men from Yazoo City, McClintock, Dixon, and Alexander would have conducted them on a comprehensive evaluation and tour of the *Hunley*.

While Singer was in Richmond demonstrating the new torpedo, Fretwell could have been placed in charge of the Mobile operation, at which time he approached various military authorities with a request for the twenty thousand–dollar bounty on destruction of the Federal gunboat *Baron DeKalb*.[15] It is not known if the Confederate government ever paid this promised bounty.

The fall of Vicksburg and the loss of the Mississippi River renewed fears of invasion throughout the coastal regions of Texas, so General Magruder called for soldier volunteers to reinforce coastal fortifications. He promised to pay each man an additional thirty dollars per month and an extra half daily ration for his services and put his engineering officers to work on fortifications from Corpus Christi to the Louisiana border.[16]

It became painfully clear to members of the Singer-Fretwell organization that they were all on the wrong side of the Mississippi River. Edgar Singer and J. D. Braman wrote personal letters expressing concern about the well-being of their families, cut off and alone in Texas (a feeling undoubtedly shared by all Texans fighting in the eastern theater).[17] Although Texas and regions west of the river were still under the jurisdiction of the Confederate government, communications between the separated territories were practically nonexistent for the remainder of the war.

In Richmond, news of the recent defeat at Gettysburg and loss of Vicksburg and the Mississippi River was causing a minor panic throughout the city. As the devastating details of General Lee's defeat in the fields of Pennsylvania continued to trickle southward into Virginia, those who had accompanied Singer from Port Lavaca prepared one of their unique underwater contact mines for inspection by a board of competent engineers. Just ten days after Gen. Robert E. Lee began his slow retreat back into war-ravaged Virginia, Confederate engineering officers witnessed a demonstration of one of Edger Singer's torpedoes in the muddy waters of the James River and filed the following report with Col. Jeremy Gilmer, chief of the Engineering Department:

> Engineer Department. July 14, 1863. Colonel: In accordance with your order of the 13th, appointing the undersigned a commission to examine and report on the merits of Mr. E. C. Singer's torpedo, we beg to state that we have carefully examined the same and submit the following report:
>
> First. "As to the plan for exploding the charge." In this plan, or lock, in our opinion, consists the great merit of the invention. The lock is simple, strong, and not liable at any time to be out of order, and as the caps which ignite the charge are placed within the powder magazine they are not likely to be effected by moisture. . . . Second. The certainty of action depends, of course, upon contact, but by the peculiar and excellent arrangement of the lock and plan of percussion mentioned above the certainty of explosion is almost absolute. . . . Third. The efficiency of its explosion, if made in deep channel, can not well be ascertained without experiment, but would be the same as submarines fired by any other contrivance. . . . We are so well satisfied with the merits of Mr. Singer's torpedo that we recommend the Engineer Department to give it a thorough test, and, if practicable, to have some of them placed at an early day in some of the river approaches of Richmond. Respectfully submitted, W. H. Stevens.[18]

Within a week after receiving the report, Colonel Gilmer forwarded a copy to the Confederate secretary of war and ordered six Singer torpedoes to be "constructed for the Richmond defenses without delay" and deployed for "use in the James River."[19] With a contract presumably signed, Singer placed James Jones in charge of torpedo production in Richmond (Robert W. Dunn and C. E. Frary also remained in Richmond until mid-August),[20] and boarded a southbound train for the recently established group headquarters in Mobile.

Singer and his associates presumably reoccupied their rooms at Mobile's Battle House Hotel (Horace Hunley is known to have had a room there, so the hotel probably was the residence of all group members then operating in Mobile).[21] Within days of Singer's return, the *Hunley* was launched in front of a cheering crowd, and another payment of twelve thousand dollars for eighty more Singer-Fretwell torpedoes was forthcoming from Mobile military headquarters.[22]

With a successful launch and underwater test conducted by McClintock and Dixon, several members of Singer's organization approached Admiral Buchanan and General Maury with an invitation to attend an underwater demonstration of the *Hunley*.

On the morning of July 31, 1863, members of the Singer Secret Service Corps towed an old, worm-eaten coal barge to the middle of the Mobile River and anchored it. Onshore military officers from both the army and navy had assembled to witness the destructive capabilities of Singer's new diving machine.[23] While the restless officers talked among themselves on the muddy shore of the river, the *Hunley*'s volunteer crew took turns squeezing through the submarine's narrow hatches several hundred yards upriver. When all the crew had reached their stations, the skipper (most likely McClintock or Dixon) ordered the narrow hatches sealed and the sea valves opened to allow water to enter the forward and aft ballast compartments. As water splashed into both tanks, the helmsman watched the water outside rise around the hull through small glass view ports that pierced the conning tower. When the water had risen to the lower lip of the forward viewports, he lit a candle next to the compass and ordered the open sea valves to be immediately closed and the propeller shaft turned.

As the tiny submarine approached the stationary coal barge, the Confederate military men on the beach quickly took notice and turned a curious eye toward the semi-submerged craft slowly approaching the barge. At the end of a long line that trailed behind the strange-looking craft was a small powder-filled cylinder bristling with contact detonators rigged to explode at the slightest touch. As the barge grew larger through the submarine's two forward viewports, the skipper firmly pushed on the heavy iron lever that depressed the outside diving planes. In an instant the *Hunley* gracefully slipped beneath the surface and vanished before the eyes of the assembled crowd.

As shouts of encouragement and delight echoed along the banks of the river, the skipper watched the depth-monitoring column of mercury rise within its glass tube and leveled the submarine off when the *Hunley* reached twenty feet. For the next couple of minutes nothing could be heard but the echo of grinding gears and the turning of the long propeller shaft as the skipper gently manipulated the rudder while watching the compass needle swing slowly from side to side. Suddenly a huge concussion enveloped the submerged vessel, causing the submarine to shudder violently and list to one side. When the shock wave had passed, both ballast compartments were quickly pumped dry, causing the submarine to rise to the surface on the far side of the sinking coal barge. To those assembled onshore, the *Hunley* had indeed proven itself a worthy adversary to the enemy's blockading fleet.

Several eyewitness accounts of these underwater tests have recently come to light. Much can be learned from these handwritten testimonials, which McClintock gathered after the war as proof of success. General James Slaughter provides this account:

> In company with Admiral Buchanan and many officers of the C.S. Navy and Army, I witnessed her [the *Hunley*'s] operations in the river and harbor of Mobile. I saw her pass under a large raft of lumber towing a torpedo behind her which destroyed the raft. She appeared three or four hundred yards beyond the raft and so far as I could judge she behaved as well under water as above it. I will add that I witnessed her experiments more than a dozen times with equal satisfaction.[24]

The destructive potential of the small submarine was obvious, so military commanders in Mobile unanimously agreed that the *Hunley* should be put into service as quickly as possible. Because Mobile Bay had relatively shallow water and strong harbor defenses, the commanders chose Charleston as the *Hunley*'s future base of operations.[25] For several weeks Charleston's outer defenses had been subjected to daily bombardments from the Federal fleet, and military strategists knew that if Charleston fell, only one other Atlantic port (Wilmington, North Carolina) would remain open to the blockade runners.

Within hours of the test, Gen. John Slaughter penned a letter of introduction for Gus Whitney and Baxter Watson. After a presumed final briefing with Captain Singer, the two agents boarded a Charleston-bound train with the following letter:[26]

> Mobile, Alabama. July 31, 1863. My Dear General [Beauregard]: This will be handed you by Messrs. B. Watson and B. A. Whitney, the inventors of a Submarine boat which they desire to submit to you for examination, and if it meets your approval to test it's [*sic*] usefulness in Charleston Harbor. So far as I am able to judge I can see no reason why it should not answer all our sanguine expectations. Nothing appears to me wanting but cool and determined men to manage it, but you will see and judge for yourself. Your Old Friend, John E. Slaughter.[27]

On the following day Admiral Buchanan sent a similar letter of recommendation to Charleston's naval commander, John Tucker:

Naval Commandant's Office Mobile, Ala. August 1st, 1863. Sir: I yesterday witnessed the destruction of a lighter or coal flat in the Mobile River by a torpedo which was placed under it by a submarine iron boat, the invention of Messrs. Whitney and McClintock; Messrs. Watson and Whitney visit Charleston for the purpose of consulting General Beauregard and yourself to ascertain whether you will try it, they will explain all its advantages, and if it can operate in smooth water where the current is not strong as was the case yesterday. I can recommend it to your favorable consideration, it can be propelled about four knots per hour, to judge from the experiment of yesterday. I am fully satisfied it can be used successfully in blowing-up one or more of the enemy's ironclads in your harbor. Do me the favor to show this to General Beauregard with my regards. Very Respectfully, Franklin Buchanan Admiral CSN.[28]

Word of the *Hunley*'s existence had already reached General Beauregard prior to the dispatching of Whitney and Watson to Charleston. As noted previously, Capt. Isaac Brown had worked with John Fretwell for some time in Yazoo City and over the ensuing months had been made well aware of the submarine being constructed in Mobile.

When the naval yards were burned and Yazoo City abandoned, Captain Brown and several of his officers made their way to Selma to await orders from Richmond. In late July Brown received the following directive from the Confederate Navy Department regarding his new command and duty station: "Confederate States Navy Department. Richmond, 29th July, 1863. Sir: Proceed to Charleston S.C. with all dispatch and report to Flag Officer Ingraham for the command of the ironclad gunboat 'Charleston' taking with you the officers and men in your command."[29]

Soon after arrival in Charleston, Brown apparently met with General Beauregard and was briefed on the current situation. Apparently during that initial meeting or shortly thereafter Brown informed General Beauregard about the submarine vessel. On August 2, 1863, just forty-eight hours after the Singer group had demonstrated the attack capabilities of their submarine to Mobile's military commanders, General Beauregard sent the following inquiry to General Maury in Alabama: "General Maury: Commander Brown C.S.N. informs me Mr. Fretwell has a submarine boat which could be used here successfully. If so please order, with consent of owners, transportation for it and party to work it."[30]

General Beauregard did not know that the two Singer operatives had already been dispatched to Charleston to seek a meeting with him and propose deploying the submarine in Charleston Harbor. General Maury's response to inform Beauregard about the two operatives is unknown because no log books recording incoming telegrams are known to exist for the Department of South Carolina, Georgia, and Florida for the period in question. General

Beauregard did meet with Whitney and Watson soon after their arrival and sent the following telegram: "August 5, 1863, General Maury: Have seen Whitney and Watson [stop] have accepted their submarine boat [stop] Please assist them to get it here as soon as possible. Signed P. G. T. Beauregard, General Commanding."[31]

With the *Hunley* firmly tied down to two flat cars and probably covered with a canvas tarp to ensure secrecy, McClintock and several members of Singer's torpedo organization headed for their new base of operations in South Carolina.[32] Although Lt. George E. Dixon is thought not to have accompanied the submarine to Charleston at this time, his company commander, Capt. James J. Williams, wrote to his wife and indicates that Dixon may have been a member of this first crew: "August 7, 1863. Dear Lizzie: I have heard that the sub-marine is off for Charleston, I suppose that Dixon went with it. With favorable circumstances it will succeed, and I hope to hear a report of its success before this month is out; still there are so many things which may ruin the enterprise, that I am not so sanguine of its triumph as Dixon."[33] This letter may prove that military authorities officially attached Dixon to the project sometime before early August.

On the morning of August 12, 1863, the locomotive that had hauled the *Hunley* and its crew from Alabama slowly steamed into the busy Charleston railroad station. For three or four days (surviving evidence seems to indicate that the *Hunley* left Mobile on August 7 or 8),[34] the tarp-covered submarine boat had traveled across the southern regions of the Confederacy, turning heads of curious citizens as it made its way to the city that had witnessed the war's opening shots. The group was met with scenes far worse than those witnessed in Mobile, because Charleston and its defending garrison were under heavy siege. Wagons pulled by straining horses, piled high with military supplies and artillery projectiles, slowly rumbled through the streets, while the booming of enemy artillery echoed in the distance.

Charleston's outer fortifications consisted of Fort Sumter, located at the mouth of the harbor with some seventy cannons, and Fort Moultrie at the southern tip of Sullivan's Island with about forty pieces of artillery.[35] In the channel between the two forts, Gen. Gabriel Rains and members of his Torpedo Bureau had anchored a string of underwater keg torpedoes to keep the enemy's ironclad monitors at bay. Guarding the southern entrance to the harbor on the northern shore of Morris Island was Battery Wagner, a hastily constructed sand fortification that had been under siege for several weeks by Federal infantry regiments that had captured the southern end of the island. It was plain to all who witnessed the siege of Charleston, known as the cradle of secession, that the tide of war was now turning against the ill-equipped and isolated Confederacy.

General Beauregard ordered the Engineering Department to unload the vessel and transfer it to a mooring in the nearby harbor without delay. While McClintock, Watson, and several Singer operatives oversaw the unloading, news of the new secret weapon spread quickly through the streets of the war-weary city. Soon a curious assortment of onlookers had assembled around the mysterious vessel and gawked as men and horses strained to lift the heavy submarine from the two flat cars. When the craft finally swung free and rolled through the streets of the city, the astonished citizens cheered with delight as this new threat to the hated Federal fleet slowly lumbered past.

The presence of the unique diving machine raised morale throughout his command, so General Beauregard, anxious to get the submarine into action as quickly as possible, issued the following order within hours after its arrival: "August 12th 1863. Major Hutson Chief Quarter Master, Department of S.C., Ga., & Fla. You will furnish Mr. B. A. Whitney on his requisition with such articles as he may need for placing his submarine vessel in condition for service. His requisitions will be approved subsequently at this office."[36]

The fact that the *Hunley* is referred to in the above order as "his submarine" (Whitney's) indicates that Whitney was chief spokesman for the Singer group during this period. It seems that most everyone in the organization quickly made arrangements to join the submarine venture in Charleston, because on August 10, Robert Dunn and John D. Braman received orders in Richmond to report to Charleston. Robert Dunn's order read, "Confederate States of America, War Department, Engineer Bureau, Richmond, Va. August 10, 1863. R. W. Dunn, Private in Shea's Battery of Artillery 'Texas,' having been detached for special service, is authorized by the Secretary of War to proceed from Richmond, Virginia, to Charleston, South Carolina, and report temporarily to General Beauregard commanding. J. F. Gilmer, Colonel of Engineers and Chief of Bureau."[37]

Some days later, C. E. Frary, still in Richmond with James Jones to help fabricate Singer torpedoes ordered by Colonel Gilmer "to be constructed for the Richmond defenses without delay,"[38] received a similar order directing him to proceed to Charleston and "report temporarily to General Beauregard."[39] Singer arrived in Charleston a couple of days later to help construct the torpedoes that would be used against the blockading enemy fleet. Horace Hunley and other members of Singer's group had arrived in Charleston by mid-August. The arrival of Hunley brought the total of Singer operatives present in Charleston to eight (of those who accompanied the submarine, only Jeremiah Donovan, an eighteen-year-old machinist's assistant at Park and Lyons, can be confirmed).[40]

Soon after the arrival of Singer and his operatives, the following directive was sent to the commander of the city's arsenal: "Charleston, S.C. August 16th 1863. Major: The commanding General desires that you will render every assistance of material and labor to Messrs. Whitney and Watson, in the construction of torpedoes to be used with their submarine vessel which he regards as the most formidable engine of war for the defense of Charleston now at his disposition."[41]

All we know about the first few days of operations in Charleston Harbor can be summed up in one or two surviving orders and dispatches. An ex-Confederate naval officer wrote many years later about the secrecy surrounding the *Hunley*: "Absolute secrecy was essential to their success, and hence scant mention of them appears in the official communications, and none whatever in the newspapers of the time."[42] Although secrecy must have been regarded as essential to the *Hunley*'s success, several contemporary letters written by members of the city's civilian population have turned up in recent years, noting that the submarine was regarded as something of a novelty around Charleston and that the public was able to view practice dives and surface operations.

Charleston native Harriet Middleton wrote to her sister on August 17, 1863. She knows that a functioning submarine is then in Charleston, is aware of the bounty offered for sinking enemy ships, and knows how long the vessel can remain submerged without injury to the crew: "If the submarine boat proves successful we may soon demolish the Yankee fleet. A man from New Orleans [most likely James McClintock] has arrived with one and the Trenholm house have offered $100,000 if he blows up the Ironsides and $50,000 for every Monitor. He brought letters from Com. Maury to Beauregard. The Government is to give one half of the value of every vessel he destroys. It was tested in Mobile on an old vessel with most satisfactory results. The boat is 40 feet long and can remain underwater one hour and ½. We are all full of hope and hear it is soon to make an attack. May it be successful."[43]

On August 18 Commissioner of Patents Rufus Rhodes sent Singer yet another letter regarding his request for a Confederate patent on his new underwater contact mine, although Captain Singer probably did not see this communication until he returned to Richmond some weeks later: "There is nothing new in placing a lock [the spring-loaded detonating mechanism] outside of the case where the explosive material is confined. The lock of every ordinary musket is outside of the chamber which confines the charge of powder. All you can claim as your invention are the mechanical elements of which the lock is composed, and the nature of the mechanism itself indicates that it is to be placed outside of the can to produce the effect."[44]

Commissioner Rhodes's letter also discusses the issue of an additional payment of twenty dollars to be forthcoming from Singer and requests additional information on minor points found on the diagrams that accompanied the patent application. Captain Singer was eventually granted a patent on his new invention, but not before a further meeting with Commissioner Rhodes and the lengthy paperwork was completed.

Sometime during this period Singer and his associates were also developing a modified version of their torpedo to be used by Confederate saboteurs against Union trains.[45] The Singer-Fretwell torpedo diagram makes it quite easy to imagine how the device could be modified for use against enemy trains. The four-pronged detonating rod that triggered the explosive could simply be replaced with one measuring about one foot long. The torpedo casing could be buried beneath a railroad track, with the detonating rod pointing upward and connected by a thin trip wire to a similar rod or stake placed on the far side of the tracks. When a passing train hit the trip wire, the detonating rod would be pulled free and the torpedo exploded. Singer agents used this supposed arrangement several times behind enemy lines.

On August 21 Hunley filed a requisition for nine Confederate uniforms for the Singer crew to wear during their nocturnal patrols. If they were captured, the Confederate uniforms could make the difference between their becoming prisoners of war or being hung for involvement with a vessel not recognized as a weapon for use in civilized warfare: "August 21, 1863. Special Requisition: For nine gray jackets, three to be trimmed in gold braid. Circumstances: That the men for whom they are ordered are on Special Secret Service and that it is necessary that they be clothed in the Confederate Army uniform. Signed H. L. Hunley, Captain."[46]

Although this document is rather brief, it is conclusive evidence that Captain Hunley and his associates had been allocated some form of military authority, for not just anyone could walk into the Charleston quartermaster's office and request Confederate uniforms (especially "three to be trimmed in gold braid"). The fact that Hunley states quite directly that "the men for whom they are ordered are on Special Secret Service" is also significant, because the men associated with the project were members of an organization whose authority could be questioned only by officers of the highest rank.

The Charleston-based ironclads, CSS *Chicora* and *Palmetto State*, acted as floating headquarters or bases of operations for the *Hunley* crew. An ex-Confederate sailor was familiar with the arrangement: "The *Palmetto State* was posted just west of Fort Johnson Wharf, and was made headquarters for the diving boat while awaiting a chance to slip out in the night and get amongst the Federal fleet."[47]

The *Palmetto State* and *Chicora* took turns steaming to the mouth of the harbor nightly to help in the defense of Fort Sumter, so the *Hunley*'s floating base changed in accordance with which vessel was then on duty, corroborated by the following message received at military headquarters on the evening of August 20: "Received by Signals From CSS Chicora, 6:25 P.M. Will be ready to assist the torpedo boat, I desire to know how far down you wish

First Torpedo Strike and Launching the *Hunley*   51

her to go. Signed Captain Tucker."[48] (Captain Tucker was probably requesting orders about how far down the harbor his vessel should tow the submarine, not how far underwater the *Hunley* should dive.)

On August 21, the morning after Captain Tucker had apparently towed the *Hunley* to the mouth of the harbor, Harriet Middleton wrote another letter to a local friend:

> You have heard of course of the wonderful fish-shaped boat, built at Mobile, and brought here in sections overland. It goes entirely underwater, has a propeller at one end, and a torpedo at the other. It has fins with valves to them to let in air, although it holds a sufficient supply to last eight men three hours after it is submerged. Papa has seen a man, who saw a man, who made a voyage in the contrivance from one wharf to another, he was twenty minutes underwater and suffered no inconvenience. The inventor tells Mr. Read he is sure the boat will do its part, and if he had not been confident that his heart would not fail him, he would never have come so far to make the attempt. They are only waiting for good weather, any bright morning we may hear that the "Ironsides" is sunk [the USS *Ironsides* was the most powerful ironclad in the blockading fleet off Charleston]. Mr. Wagner offers $100,000 for her destruction, and 50,000 for the "Wabash" [the flagship of the blockading squadron]. I hear Beauregard says if he can only get rid of the "Ironsides" he thinks he can manage the Monitors.[49]

The apparent secrecy that surrounded the entire *Hunley* operation makes the recent discovery of these Middleton letters very significant. Although they could be viewed as being based on nothing more than local rumors, without them we would be left with only a handful of dry dispatches and vague postwar accounts regarding the early days of the submarine's deployment. Harriet Middleton was very familiar with the everyday goings-on of the operation, and her knowledge is comprehensive, for she sheds light on several aspects of the venture that remain somewhat puzzling today.

On the very night that Middleton sent her letter, the sleeping citizens of Charleston were rudely awakened by the sound of a terrific explosion. On Morris Island, some four miles distant, a Federal artillery regiment had managed to mount a huge siege cannon in the swamps, aimed not at Charleston's outer defenses but at the very heart of the city. At 1:30 a.m. the first shot was fired into Charleston, followed by over thirty more exploding projectiles before the sun rose. The huge two hundred–pound Parrott rifle that had rained destruction down on Charleston, would soon come to be known throughout the city as "The Swamp Angel."[50]

On August 22, the day following Charleston's night of terror, the editor of the *Daily Courier* wrote a blistering editorial titled "The Bombardment." The opening paragraph read, "The startling events that have occurred since our last issue have opened up a new chapter in the history of the war. Our ferocious foe, maddened to desperation at the heroic obstinacy and resistance of his powerful combination of land and naval forces to reduce Fort Sumter and our batteries on Morris Island, tries the horrible and brutal resort, without the usual notice, of firing, at midnight upon the city, full of sleeping women and children, to intimidate our commanding general into a surrender of those fortifications. Our people are nerved for the crisis and with calm determination have resolved on making it a struggle for life or death."

With the city's civilian population literally under the guns of the enemy, a bold strike from the *Hunley* at the heart of the enemy fleet was needed more than ever. As the military situation in Charleston grew increasingly desperate, General Beauregard and Flag Officer Tucker became impatient with the owners of the submarine. If its untested crew were unable to make a successful attack on the blockading fleet, perhaps a military crew under the command of a combat-experienced naval officer could.[51]

In the early hours of August 23, a situation developed that may have affected Beauregard's decision to seize the *Hunley* from its owners. At about three o'clock that morning a small group of Federal ironclads approached the harbor's entrance and commenced shelling Fort Sumter. As Forts Sumter and nearby Moultrie returned fire, a thick fog rolled in from the Atlantic, rendering targets invisible to both defender and attacker alike. When the fog lifted, it was discovered that one of the attacking monitors had run aground on a hidden sandbar about a thousand yards in front of Fort Moultrie's batteries. As the huge naval guns within the fort started to target the grounded ironclad, fog once again rolled in, causing the target to again vanish from sight. In a report concerning the unusual event, a disgusted Gen. Thomas Clingman (military commander of Sullivan's Island) wrote that the grounded ironclad could have been sunk if only it had been in view for another thirty minutes.[52]

With the rare opportunity of destroying one of the enemy's most powerful ships botched, Clingman would have been in no mood to hear excuses from the *Hunley*'s assumed commander James McClintock. Within hours after writing his report on the failed attempt to destroy the grounded monitor, Clingman sent a short note to his aide, Captain Nance: "The torpedo boat started at sunset but returned as they state because of an accident. Whitney says that though McClintock is timid, yet it shall go tonight unless the weather is bad."[53] The *Hunley* was perhaps moored behind Fort Moultrie (in an area known as the cove) during daylight hours; otherwise, it seems highly unlikely that the general would have had anything to do with the operation.

The McClintock crew (probably Gus Whitney, Robert Dunn, Baxter Watson, John D. Braman, Jeremiah Donivan, C. E. Frary, and perhaps E. C. Singer himself)[54] made at least three nocturnal attempts against the

Federal monitors anchored at the harbor's entrance. General Clingman regarded these as ineffective because he sent Captain Nance another unflattering message concerning the conduct of the *Hunley*'s crew: "The torpedo boat has not gone out, I do not think it will render any service under its present management."[55] Within twenty-four hours, the Charleston military seized the *Hunley* and turned it over to the Confederate navy.[56]

A call for naval volunteers to crew the mysterious contraption was quickly sent throughout the Charleston squadron. Within hours, Lt. John A. Payne, a combat-experienced officer then attached to the ironclad CSS *Chicora*, stepped forward to request command of the *Hunley*.[57] The following directive written some forty-eight hours after the navy took control of the submarine indicates that some of the Singer group were apparently still attached to the venture in an advisory capacity: "August 26th, 1863, General: The bearer C. L. Sprague has come recommended as one ingenious in matters relating to submarine torpedoes and is directed to report to you to be attached to the submarine vessel of Whitney and company. He may be of service to the Naval officer who has volunteered to take that vessel in hand, and it were well to place them in communication as also with Mr. Whitney."[58] However, Gus Whitney, the Singer Secret Service Corps' chief spokesman, "died in [late August] 1863, as the result of pneumonia contracted while experimenting with the boat."[59] Thus, Whitney could be considered the first casualty of the vessel soon to be referred to as the "Peripatetic Coffin" throughout the Charleston squadron.[60]

Throughout this period of redeployment and acquisition of new volunteers to crew the *Hunley*, Fort Sumter was subjected to a fierce, and constant, Federal artillery barrage from both land and sea. Hundreds of shells struck the fort daily, and fear that the garrison would soon be forced to abandon Charleston's main defensive structure quickly became a source of great concern at Beauregard's headquarters. If Sumter fell, the city would soon follow, for the harbor would then be effectively sealed off to the ever-dwindling number of blockade runners.

On the same day that Charles L. Sprague was assigned to the *Hunley* to act as torpedo engineer, the following stern directive was sent from Charleston military headquarters to the besieged garrison of Fort Sumter, then isolated and alone at the harbor entrance: "August 26, 1863, General: Fort Sumter must be held to the last extremity and not surrendered until it becomes impossible to hold it longer without an unnecessary sacrifice of human life. Evacuation of the fort must not be contemplated one instant without positive orders from these headquarters."[61]

About forty-eight hours later, Harriet Middleton was again penning a letter to a local friend concerning talk throughout Charleston that the Singer crew had apparently been successful in passing under the most powerful ironclad then in the blockading fleet. Much of the letter, however, must be regarded as local rumors.

> August 28, 1863. The fish-boat has been under the "Ironsides," but the torpedo got off, and the water was too shallow where she lay for them to have done much injury. They must have sixty feet of water to dive into, in order to get force enough in rising to make a hole in her bottom. People are offering the eight men who go in the machine $10,000 each for their places and chances in the enterprise. Some say the thing looks like a log on the water, others that it is like a gigantic metallic coffin, but all say it is wonderful, only the sanguine believe it will work wonders.[62]

Ex-Confederate naval lieutenant C. L. Stanton provided eyewitness accounts regarding his experiences with the *Hunley* after the government took command of it:

> The first submarine craft in the world worthy of the name, so far as I know... was the fish boat, which operated in Charleston Harbor. John A. Payne, her first commander, was a lieutenant on the Chicora, and her first volunteer crew were seamen from the vessel, nine in all including the commanding officer.... It was built of boiler iron about thirty feet in length, five feet in diameter, had movable flanges, or fins, on the side, and had a propeller that was worked by hand from the inside. These fins or wings, were depressed for diving and elevated when it was desired to come to the surface. There were two manheads, one forward and the other aft, through entrance was had to the interior....
>
> One day when Lieutenant Payne, my friend and shipmate, was aboard the Chicora I arranged to go down under the water with him; but as the boat was obliged to leave before my watch on deck was over, Lieutenant Charles H. Hasker took my place. She dived about the harbor successfully for an hour or two and finally went over to Fort Johnson, where the little steamer Etiwan was lying alongside the wharf. She fastened to her side with a light line with the fins in position for diving. Both manheads were open.
>
> Payne was standing on top of the fish boat, and Lieutenant Hasker's body was half way through the manhole, when the steamer started to move away from the wharf. The line by which the boat was attached to the steamer snapped, and she went to the bottom in five fathoms of water like a lump of lead. Payne escaped by jumping aboard the Etiwan, while Hasker was carried down with the boat, the manhead closing down as he struggled through the opening and crushing his foot. He told me afterwards that he had to stoop down and raise the manhead to release the imprisoned foot. He came to the surface in an exhausted condition and was picked up by one of the Chicora's boats. The crew of seven men were drowned.[63]

Charles Hasker provides additional details:

We were lying astern of the steamer Etewan at the wharf near Fort Johnson, with the manholes open. Payne gave the order to go ahead, sheared the boat off from the steamer before he cast off the bowline. The boat careened, and commenced to fill, Lieutenant Payne was fouled by the bowline, but got himself clear. Two men got out the after manholes. The rest of them went to the bottom. Five were drowned. It was a critical time for me. I was next to Payne at the forward end of the boat, just aft the shaft which controlled the lateral fins. I had to get over this shaft and out of the manhole. The boat was nearly filled with water.

I thought it was useless to try and force myself through the column of water coming through the manhole, but thought of my wife and children, and breathed a prayer. It was only three words "God help me!" God heard and answered my prayer. I took hold of the rim of the manhole, and drew myself up through the water, when the manhole cover fell on my shoulder. I now worked my body out, was caught by my left leg, the calf of which was pressed in two by the pressure of the manhole plate. When the boat touched bottom I felt the weight relax, stooped and lifted the plate from my leg, and drew it out, and swam for the surface. I was picked up by midshipman Daniel Lee [a nephew of Gen. R. E. Lee] and taken on board my vessel, the ironclad Chicora, more dead than alive.[64]

Word of the disaster traveled quickly throughout Charleston's civilian population. On the very day that the *Hunley* sank and five crew drowned, Harriet Middleton was already discussing the event in a letter to a friend: "It seems a pity that the fish-boat should have been turned over to the Government, we might have had a better chance at the 'Ironsides' if she had been bought by Trenholm and taken out by Jefferson Bennett, as was first proposed. It is too bad to have her lying at the bottom of the bay, when so many long-headed men who understood machinery saw no reason why she should not succeed in sending the enemy down."[65]

The *Charleston Daily Courier* printed the following account:

Unfortunate Accident—On Saturday last while Lieutenants Payne and Hasker, of the Confederate Navy, were about experimenting with a boat in the harbor, she parted from her moorings and became suddenly submerged, carrying down with her five seamen who were drowned. The boat and bodies had not been recovered up to a late hour on Sunday. Four of the men belonged to the gunboat Chicora, and were named Frank Doyle, John Kelly, Michael Cane and Nicholas Davis. The fifth man, whose name we did not learn, was attached to the Palmetto State.[66]

Of all the high-ranking military officers who interacted with Captain Singer and members of his Secret Service Corps, none was more knowledgeable about the subject of underwater explosives than Gen. Gabriel Rains, commander of the Confederate army's Torpedo Bureau. When the *Hunley* arrived from Mobile, General Rains and his small staff of torpedo engineers had for some months been manufacturing contact-sensitive keg torpedoes (fabricated from old beer kegs modified with large wooden cones attached at both ends) at a facility at the foot of Broad Street.[67]

The army, not the navy, was in charge of torpedo operations throughout the region, and the duties extended to both manufacture and deployment. While General Rains oversaw the overall operation (his official duties as head of the Torpedo Bureau extended to tours and inspections of operations throughout the southeast), his immediate subordinate, Capt. M. M. Gray, was delegated the responsibility of gathering materials and arranging deployment.[68]

The Singer group may well have been placed in communication with General Rains upon arrival, because Fretwell and Singer both interacted with Rains and his department several times during the war (Fretwell's name does not appear in any known Charleston order, requisition, or dispatch, so he may have stayed in Alabama to oversee torpedo operations in Mobile Bay). Rains was well aware of the *Hunley* and its capabilities. Rains wrote in detail about the *Hunley* and its history, including these observations:

> The cigar boat of Mobile was a submarine boat about 35 feet long made somewhat in the shape indicated, with capacity sufficient to hold nine operators. In the rear of this boat was her propeller. This boat was constructed of boiler plate iron with manholes at the top with iron trapdoors for egress, having two fins, one on each side near the middle, and steering appendages controlled within. Her crew consisted of eight men to work the propeller by means of cranks, one to steer. The speed was about four knots. She could remain submerged ½ hour. She was to drag by rope a floating torpedo under a vessel's hull. This boat could go below the surface and come up again to the top, but was defective in obtaining proper fresh air for the respiration of the crew. She was controllable in smooth but not rough water. . . . Her efforts in Charleston Harbor were attended by the drowning of a Negro, then five men in her by being capsized by a steamer's swell with the covers of the manholes open. . . . This boat was brought to the wharf where we were preparing torpedoes, but on account of the mishaps, I would have nothing to do with her.[69]

With the loss of the recently arrived submarine and five crew, the excitement once felt throughout the war-weary city was dashed. If the *Hunley* remained abandoned in the mud at the bottom of the harbor, little could be done to thwart the Federal fleet's stranglehold on Charleston. Within seventy-two hours General Beauregard, anxious to learn why his new secret weapon was lost, sent inquiries throughout his command.[70]

He then dispatched the following telegram to the commander of the first military district on the other side of Charleston Harbor: "General Ripley: Fish Torpedo still at bottom of bay, no one working on it. Adopt immediate measures to have it raised at once. Put proper person in charge of the work. Inform Lieutenant Payne of my orders. General Beauregard."[71] With no other viable offensive weapon then at their command, General Beauregard and Flag Officer Tucker apparently decided that the only course of action was to salvage the submarine, find another volunteer crew, and put the *Hunley* back into service.

The extraordinary task of recovering the forty-foot submarine from the murky depths of Charleston Harbor was given to civilian divers Angus Smith and David Broadfoot, two Scottish immigrants who had been of service to Confederate authorities several times before.[72] The two men's main task was assisting in the nocturnal deployment and anchoring of the keg torpedoes at the mouth of the harbor. To be a deep-sea diver in 1863 was quite unique, for diving technology during the mid-nineteenth century was still in its infancy with very little known about the deadly bends or other hazards associated with breathing air under pressure.

For several days after the sinking, Smith and Broadfoot groped around the dark bottom of the harbor in their copper diving helmets, attaching chains and ropes to the hull of the silent submarine.[73] We do not know how many members of the Singer Secret Service Corps were present at the recovery operations or how many actually stayed on in Charleston to act as advisers. Robert Dunn and J. D. Braman in all likelihood remained in Charleston after the death of their friend Gus Whitney. E. C. Singer returned to Richmond to continue meetings with government officials shortly after the *Hunley* was turned over to the navy.

Horace Hunley had returned to Mobile by September 12, where he is registered at the Battle House Hotel.[74] He was actively recruiting another crew from the mechanics and machinists who had helped construct the vessel. But Hunley journeyed back to Charleston within days of his arrival in Alabama to try to take back command of the vessel from General Beauregard with "a crew from Mobile who are well acquainted with its management."[75]

While Horace Hunley lingered in Mobile to recruit a replacement crew to take back to Charleston, Smith and Broadfoot were successful in raising the submarine. On September 14, General Ripley sent the following dispatch to General Beauregard's chief of staff, Gen. Thomas Jordan: "Headquarters first military district Charleston, Sept. 14, 1863. General: I have the honor to inform you that the torpedo submarine boat was brought up to the city this afternoon and is in the vicinity of the RR wharf, in charge of Lt. Payne C.S.N. As I shall be absent tomorrow, I have directed Major Pringle to give any assistance in fitting her for service which he shall be called upon for."[76]

This short report confirms that Payne was still in command of the ill-fated submarine following the salvage operation at Fort Johnson. However, the question of whether he could find another volunteer crew willing to squeeze through the narrow hatches and reman the accident-prone vessel was anyone's guess.

Around the same time that the late crew of the *Hunley* was being removed and hastily buried at Mariners Grave Yard,[77] Captain Singer (with perhaps James Jones and others who were then with him in Richmond) was probably meeting with Rufus Rhodes in regard to questions about a patent raised in Rhodes's letter to Singer dated August 18. On September 15, 1863, Capt. Edgar C. Singer received an official Confederate patent for his revolutionary contact mine. With this document, Singer and all those associated with him would henceforth be protected under Confederate law against anyone who might infringe on their design by fabricating a similar device.[78]

With Captain Singer then in Richmond, and Horace Hunley presumably trying to recruit a replacement crew in Mobile, Robert W. Dunn and John D. Braman (and perhaps James McClintock and Baxter Watson, for no records as to their whereabouts in early September 1863 have thus far come to light) were taking a keen interest in a semi-submersible, cigar-shaped, wooden torpedo boat known as the *David*.

Singer was quite pleased with Dunn and Braman's idea to fabricate inexpensive wooden torpedo boats on the plan of the *David*, and he spent much time and effort in Richmond trying to secure funding for the venture from the secretary of war and Confederate Congress. In the months to come Braman expanded the simple design of the *David* into an armored vessel that Singer himself confidently stated in a letter home "promises the redemption [recapture] of New Orleans and the Mississippi River."[79]

While Braman and Dunn went over the diagrams of their new torpedo boat, Hunley and Dixon were on their way to Charleston from Mobile. Determined to prove the submarine worthy of General Beauregard's confidence, Captain Hunley sent the following request for command of the vessel to military headquarters soon after his arrival:

> Charleston, September 19th, 1863. General G. T. Beauregard. Sir: I am part owner of the torpedo boat the "Hunley." I have been interested in building this description of boat since the beginning of the war, and furnished the means entirely of building the predecessor of this boat, which was lost in an attempt to blow up a Federal vessel off Fort Morgan in Mobile Harbor. I feel therefore a deep interest in its success. I propose if you will place the boat in my hands to furnish a crew (in whole or in part) from Mobile who are well acquainted with its management and make the attempt to destroy a vessel of the enemy as early as practicable. Very respectfully your servant, H. L. Hunley.[80]

General Beauregard understood that Lt. George Dixon had been in charge of the submarine during trials in the Mobile River and had demonstrated its attack capabilities to Admiral Buchanan. Writing from memory over a decade after the conflict had ended, Beauregard stated, "After the recovery of the sunken boat Mr. Hunley came from Mobile, bringing with him Lieutenant Dixon, of the Alabama Volunteers, who had successfully experimented with the boat in the Harbor of Mobile, and under him another naval crew volunteered to work it."[81]

Within seventy-two hours after military headquarters had received Captain Hunley's offer, and presumably met with this newest Singer spokesman to discuss the transfer of the submarine back to its owners, General Jordan issued the following order to Maj. John Trezevant, commander of the Charleston Arsenal: "Charleston, Sept. 22nd, 1863. Major: The submarine Torpedo Boat has been placed in charge of Capt. H. L. Hunley, with a view to prompt repairs all dispatches essential and vital, I am instructed to request you to have all work done for Capt. Hunley that he may require with the utmost celerity and to supply such material as he will requisition as the mechanics under his control can apply. His requisitions will be approved at these Head Quarters."[82]

While Hunley and Dixon (and other Singer group operatives still in Charleston) monitored repairs to their vessel, the new Mobile crew selected by Hunley were then preparing to join their comrades in South Carolina. Lt. William Alexander describes events that took place at the Park and Lyons machine shop in late September 1863:

> General Beauregard then turned the craft over to a volunteer crew from Mobile known as the "Hunley and Parks crew." Captain Hunley and Thomas Parks, a member of the firm in whose shop the boat had been built, were in charge, with Brockband, Patterson, McHugh, Marshall, White, Beard and another as the crew. Until the day this crew left Mobile, it was understood that I was to be one of them, but at the last moment Mr. Parks prevailed on me to let him take my place. Nearly all of the men had some experience in the boat before leaving Mobile, and were well qualified to operate her.[83]

From an 1899 letter written by William Alexander, we know that the unknown crewman mentioned was undoubtedly Henry Dillingham,[84] a steamboat engineer who in all likelihood was well acquainted, or perhaps good friends, with the *Hunley*'s future skipper Lieutenant Dixon (who was also a steamboat engineer before the war).[85] Dixon was then on detached duty with the Singer group, according to his Confederate war record at the National Archives: "1st. Lt. George E. Dixon, Co. 'A' 21st Regiment Alabama Infantry detached October 1st, 1863 (30 days) by order of General Maury."[86] The optimistic Maury seems to have concluded that a month would be all that was required for the young lieutenant to accomplish his mission in Charleston.

Unfortunately, very little has been discovered regarding the background of George Dixon prior to his enlistment in the 21st Alabama Regiment. Surviving examples of his handwriting and a few remaining letters indicate that the young officer had received a thorough education, for his writings reflect a well-lettered gentleman who possessed an excellent command of the English language.

Within days after Captain Hunley had requested that the submarine be returned to the command of the Singer Secret Service Corps, the recently recruited crew of Mobile adventurers arrived at the Charleston rail station and immediately set about reacquainting themselves with the mysterious vessel that had, just weeks before, taken the lives of five of their fellow countrymen. With the Federal grip on Charleston tightening daily, everyone hoped that the little submarine could in some way overcome its tragic past.

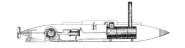

Richmond Invests in the Corps   3

WHILE CAPTAIN SINGER AND SEVERAL ASSISTANTS attended meetings in Richmond and arranged facilities in which to manufacture their recently patented contact mine, Horace Hunley and Lt. George Dixon prepared to redeploy the group's submarine in South Carolina. While the unpleasant task of cleansing the interior of the submarine was taking place at a Charleston dock, Singer operatives Robert Dunn and J. D. Braman were perhaps introducing Lieutenant Dixon and the recently arrived replacement crewman, Henry Dillingham, to the design of the *David* torpedo boat. Both Dixon and Dillingham were experienced steam engineers, so Dunn and Braman would probably have sought their opinions about the power source of the *David* since they were planning to design a similar craft. Dixon and the skipper of the *David* had become well acquainted by early October, and Dillingham and Dixon most likely had examined the interior and power source of the cigar-shaped torpedo boat.

While bold plans for an underwater attack were being hatched in South Carolina, officers of the Confederate Engineering Department in Richmond wanted proof of the destructive capabilities of Singer's torpedo since a comprehensive test of the device had not yet been carried out. At their Richmond headquarters at the Spotswood Hotel Singer and his associates made final arrangements for the torpedo demonstration for the morning of October 13.[1] Allen Pinkerton (the founder of the Union Secret Service) wrote after the war about the Spotswood Hotel becoming the main gathering place for various operatives and agents of the Confederate Secret Service.[2] In Pinkerton's 1883 book, *The Spy of the Rebellion*, he stated that one of his agents had visited the headquarters of the Confederate Secret Service, which was jokingly referred to by southern operatives as "The Subterranean Room." Pinkerton's informant reported that the operation was located in "a large room on the third floor of one of the main hotels in Richmond," and the bureau employed "nearly fifty persons, some of whom are constantly in the field carrying dispatches, gaining and bringing in information from the Yankee lines."[3] Pinkerton's description and a personal letter from Singer to his wife provide overwhelming evidence that both are describing the Spotswood Hotel.

James Jones was placed in charge of the James River torpedo demonstration and was also responsible for the Singer group's Virginia manufacturing facility. On the morning of October 13, 1863, Jones oversaw the towing of an old schooner to the middle of the James River and had it anchored like the coal barge had been moored for the *Hunley* demonstration in Mobile some two months earlier. Onshore stood Edgar Singer and several of his Virginia-based operatives who had by then manufactured more than seventy-five Singer-Fretwell torpedoes at their Richmond facility (some of which had already been deployed in Virginia waterways). James Jones wrote a letter on the afternoon of October 13 that reveals what several high-ranking officers of the Confederate Engineering Department witnessed that morning: "This morning I had an experiment for the department of which I am under contract. I set a machine [torpedo] in the James River and let a schooner float over it, and it was one of the prettiest sites I ever seen. It was a perfect success, it raised the schooner about ten feet in the air and when she came down there was nothing left but pieces. Colonel Stevens and Williams and some other smaller officers was down, and they was all much pleased."[4]

Some days prior to the demonstration Hunley and Dixon had completed repairs on the group's submarine and returned the vessel to the waters of Charleston Harbor. The exact events that occurred within the submarine's narrow hull after Captain Hunley and First Officer Thomas Parks sealed the hatches are not known. Eyewitnesses standing on the Cooper River docks and aboard the CSS *Indian Chief*, a floating barracks that housed various crew members assigned to the harbor-bound gunboats,[5] agree that they observed nothing unusual that morning. All concurred that they saw the submarine preparing to make a typical dive under the *Indian Chief*. The *Hunley* approached the vessel in the normal fashion, and its diving planes were depressed several hundred feet off the starboard side of the ship; then it disappeared beneath the surface and was not seen again. General Beauregard wrote of the tragic event some ten years later: "Lieutenant Dixon made repeated descents in the harbor of Charleston, diving under the navy receiving ship which lay at anchor there. But one day when he was absent from the city, Mr. Hunley, unfortunately, wishing to handle the boat himself, made the attempt. It was readily submerged, but did not rise again to the surface, and all on board perished from asphyxiation."[6]

C. L. Stanton remembers what he saw on that overcast morning while on temporary duty aboard the CSS *Indian Chief*: "I happened to be aboard the receiving ship Indian Chief on some temporary duty off Adger's Wharf when the fish boat was observed approaching the vessel on the starboard side, and when within a biscuit throw disappeared and successfully dived under the ship, reappearing on the port side a short distance from the docks. Presently she dived again, and when, after half an hour had elapsed, she failed to come to the surface we knew the men in her were dead."[7]

This report is recorded in the Journal of Operations kept at Confederate headquarters:

> Raining again this morning, and too hazy to get report on the fleet. An unfortunate accident occurred this morning with the submarine boat, by which Captain H. L. Hunley and 7 men lost their lives in an attempt to run under the navy receiving ship. The boat left the wharf at 9:25 a.m. and disappeared at 9:35. As soon as she sunk air bubbles were seen to rise to the surface of the water, and from this fact it is supposed the hole in the top of the boat by which the men entered was not properly closed. It was impossible at the time to make any effort to rescue the unfortunate men, as the water was some 9 fathoms deep.[8]

From the book of endorsements that summarized communications received at Charleston headquarters comes an acknowledgment of a report filed by Lieutenant Dixon several hours after the accidental sinking: "Lieutenant G. E. Dixon. Charleston, October 15, 1863. Reports the loss of the submarine torpedo boat, Captain Hunley and seven men." The exact contents of Dixon's report are unknown because the original document has long since vanished.

Even though the *Hunley* had now cost the lives of two crews, the *Hunley* had proved to be a formidable weapon of destruction in the capable hands of George Dixon. Divers Smith and Broadfoot were immediately recruited to search the chilling depths of Charleston Harbor for the missing submarine and crew.[9] On the same morning that the search operations commenced below the keel of the CSS *Indian Chief*, General Ripley issued the following orders from military headquarters: "October 16th, 1863. Major: Give transportation to Lieutenant Dixon 21st Regiment Alabama Volunteers, and Henry Dillingham to Mobile and return, on business connected with the submarine torpedo boat."[10]

Apparently both George Dixon and Henry Dillingham had not yet given up on the submarine, and they made their way to Mobile. On October 17, 1863, Flag Officer Tucker received the following request from military headquarters: "Captain Angus Smith engaged in raising the submarine torpedo boat which was unfortunately sunk a few days ago, requires the assistance of several boats and crews to endeavor to raise the vessel. I therefore request that you will give him the necessary aid in this matter, and also that the receiving ship under which the torpedo boat is thought to lie, may be moved from its present position, so as not to interfere with the operations of dragging for the boat."[11] The Charleston Harbor journal of operations includes this entry for the next day: "Oct. 18, 1863, Mr. Smith provided with submarine armor, found the sunken submarine boat today in 9 fathoms of water. The engineering department was instructed to furnish Mr. Smith all facilities in the way of ropes, chains, etc., that an attempt might be made to recover the boat."[12]

While Charleston military headquarters was trying to recover the *Hunley*, most of Tennessee was under the yoke of a Federal occupying force commanded by Gen. Ulysses S. Grant. With headquarters at Nashville, Grant and the officers attached to his Military Division of the Mississippi oversaw operations throughout the state while gathering ordnance and manpower for renewed military expeditions into additional southern regions then perceived to be in rebellion against the Federal government.[13] Grant's army required some thirty boxcars of rations daily, with ten cars being designated for the sole purpose of transporting beef cattle. General Grant sent an order to the superintendent of military railroads regarding this constant need for supplies: "The road should be run to its utmost capacity, and should there be at any time spare cars, load them with rations or forage and send them through. On no account fail to send the thirty cars daily with rations."[14]

Grant's army of occupation operated in regions hostile to the Federal government, so special orders constantly circulated among the civilian population who, for the most part, continued to support the Confederate cause:

> General Order Number 73. When a country is occupied by a military force, it is a violation of the law of war for the inhabitants to convey any information or give any aid or benefit to the enemy. Therefore it is hereby ordered that any person who conveys to the enemy any information detrimental to the United States Government, either by letter or by word, or in any other way whatever, will be treated as a spy.... Any citizen guilty of the above acts, or of giving aid or comfort to the enemy in any way, will be swiftly punished to the utmost extent of the law, his property seized and appropriated to the United States Government, and his family sent south of the Tennessee River. By order of Brig. General G. M. Dodge, Assistant Adjutant General.[15]

General Grant was well aware that a constant source of supplies and ordnance were vital to maintain his hold on occupied Tennessee. Southern commanders then plotting to retake the territory quickly hatched plans for Confederate operatives based in northern Alabama and Georgia to disrupt the Union supply trains. Singer had submitted plans to the Richmond War Department for a railroad torpedo prior to August 1,[16] and some of his operatives were dispatched to Tennessee with these explosive devices sometime in mid-October. Proof of this western deployment of Singer operatives comes from Robert Dunn's letter to General Magruder in April 1864 and several federal dispatches sent throughout Tennessee. Dunn states, "In Tennessee we have blown up eight railroad trains. These successes show conclusively the certainty of explosion of our torpedoes."[17]

Union records show that the trains were destroyed during a span of several weeks from the last days of October through early November 1863. Some of the first Federal reports record such an attack : "Decherd Tennessee, October 26, 1863. 1 a.m. Colonel McCook. Sir: Keep your command saddled all night, and stand to horse at daylight in the morning. Keep a good lookout and camp guards wide awake. The explosion was at the mountain. Rumored a torpedo blew up a locomotive. By order William H. Sinclair, Assistant Adjutant General."[18] Apparently, Singer's torpedo men had been hard at work throughout the night, for by the time the following report was filed, the destruction of a locomotive during the predawn hours of October 26 was no longer a rumor:

> Nashville Tennessee, October 27, 1863. 3:30 p.m. Major General Joseph J. Reynolds, Chief of Staff. Sir: I have the honor to report another accident upon the road last night. The rebels placed a torpedo upon the track just south of the tunnel, and as soon as the engine struck it, it blew her up, throwing her across the track and making a complete wreck of her. I have a large force hard at work getting things righted again. W. P. Innes, Colonel and Military Superintendent."[19]

The whereabouts of Singer agents L. C. Hirshberger, William Longnecker, F. M. Tucker, and C. E. Frary are not known during this time, so they may well have been part of the group of saboteurs dispatched to play havoc with General Grant's supply trains. Confederate general P. D. Roddey sent a report to Gen. Braxton Bragg that he was well aware of the railroad sabotage taking place in Tennessee, for after notifying Bragg that there was "nothing to prevent my making a successful expedition [raid] into Tennessee," he added, "I have full confidence in the detachments sent out doing great damage to the road. . . . If you will send me some torpedoes, I will put them under the track to a certainty."[20]

The Federal superintendent of military railroads filed a report indicating that Singer's men had struck quickly at several points along the line during the nights following their first success:

> Nashville, Tennessee, November 2, 1863. Major General U.S. Grant: Your dispatch received. I took charge of the road this morning. . . . Have three engines at Jefferson Indiana which can be here and ready for service by Saturday night next. Have seven engines in the shops for repairs, two of which will be out in two days. The others will require longer time. There are four engines off the track at different points. These I will get up as soon as possible. J. B. Anderson, Superintendent Military Railroads.[21]

Federal railroad personnel tasked with keeping the line open quickly replaced spans of track blown out by Singer's torpedoes.[22] The engines blown from the tracks remained abandoned until they could be retrieved. No deaths were reported in connection with these acts of sabotage, and the railroad was never closed for more than a few hours to a day at most, so this dangerous mission had little effect on the enemy's overall capacity to wage war. Gen. William T. Sherman noted that the "use of torpedoes in blowing up our cars and the road after they are in our possession, is simply malicious. It cannot alter the great problem, but simply makes trouble."[23] However, he had a rather harsh opinion of those engaged in instigating these acts of "trouble" and continued: "If torpedoes are found in the possession of an enemy to our rear, you may cause them to be put on the ground and tested by wagon loads of prisoners, or if need be, citizens engaged in their use."

Since such a harsh fate awaited anyone caught possessing torpedoes behind enemy lines, Singer's untested operatives would surely have been provided with documents stating that they were in Confederate service, for such papers are known to have been given to naval personnel on torpedo duty.[24] Singer's saboteurs planted several of their unique devices over a span of about three or four weeks during the autumn of 1863.[25] Singer's men may have been temporarily attached to one of the guerrilla "detachments sent out doing great damage to the road" or worked independently behind Grant's lines from safe houses operated by local sympathizers familiar with the terrain. The torpedoes were planted under cover of darkness along remote stretches of track far removed from established Union bases of operations. Whether the operatives arrived on horseback or on foot, they quickly planted the torpedoes by digging a pit beneath the track and then armed them.

After they placed the powder-filled container, they attached a trip wire to the spring-loaded detonating rod and stretched it across the track so it would snag the cow catcher of the first enemy locomotive that passed. After deploying perhaps several devices throughout the night at various locations along the track, the weary saboteurs would have retreated deep into the Tennessee forests before sunrise or perhaps returned to a local safe house. These acts of sabotage peaked during mid-November and then dropped off somewhat (perhaps because the agents were simply running out of torpedoes) during the first weeks of December.

The federal quartermaster in charge of supplying Grant's army sent a dispatch defining a plan to patrol the tracks against saboteurs, which was quickly implemented at military headquarters: "Nashville, Tennessee, December 23, 1863. Brigadier General Whipple: The road has failed me for the last three days, in consequence of accidents. . . . Torpedo taken up last week on the road, which fortunately did not explode, but one of our best engines thrown off the track yesterday near Decherd by rail being taken out. The road is not sufficiently patrolled, and especially between this point and Murfreesborough; men from station to station should meet every two hours. J. L. Donaldson, Quartermaster." General Whipple immediately sent the following communication to General Slocum: "Chattanooga, Tennessee. December 23, 1863. Major General Slocum: Direct patrols along the railroad to be more vigilant and meet between posts every two hours; accidents are getting numerous; one of our best engines thrown off track yesterday night near Decherd by rail being taken out. Torpedo taken up last week, which fortunately did not explode. W. D. Whipple, Assistant Adjutant General."[26]

As a result General Grant issued the following order from his headquarters in Nashville:

> Headquarters Military Division of the Mississippi, Nashville, Tennessee. December 23, 1863. Special Order Number 27. Major General Rousseau commanding District of Nashville, will make such dispositions as to have patrols of at least three men pass over each point of the railroad between this city and Murfreesborough as nearly as may be once in two hours, both by day and night. The duties of these patrols will be to arrest any person tampering with the road, to remove any obstacles and warn trains of all danger, and in case of accident to give assistance to the conductor in charge of the train. By order of Major-General U. S. Grant.[27]

Now that Union cavalry patrols were examining every section of track on an hourly basis both day and night, it became more unlikely that an explosive device could be planted and successfully detonated beneath the iron wheels of an enemy locomotive, so the railroad sabotage quickly came to an end. Although there were reports of occasional acts of destruction against the line in the months that followed,[28] most caused little more than temporary shutdowns and were generally regarded as a controllable annoyance.

While Singer's operatives buried their first torpedoes in late October, Dunn and Braman were poring over their designs for a new torpedo boat similar to the *David*. Dunn wrote to General Magruder about the proposed construction of "cigar boats" throughout the South: "While operating in the later place [Charleston] it was fully demonstrated, that torpedoes could be used on the prow, placed on the bow of a boat, without any damage to herself, yet carrying certain destruction to the vessel attacked. On the first of November last, the writer returned again to Richmond, to obtain assistance for the construction of such boats as would enable us to operate with safety to our crews, and at the same time, strike terror to the enemy."[29] We do not know the exact design of the proposed vessel, but Singer's group spent much time and effort on the subject of torpedo boats. Radical design variations from those of the *David* include (as strange as it may sound for the 1860s) incorporation of self-propelled torpedoes discharged through tubes from beneath the water's surface.[30]

On November 7, some three weeks after the *Hunley* had vanished, the slime-covered iron hull once more broke the surface of Charleston Harbor.[31] General Beauregard later described the scene when the hatches of the silent submarine were pried open: "When the boat was discovered, raised and opened the spectacle was indescribably ghastly; the unfortunate men were contorted into all kinds of horrible attitudes, some clutching candles, evidently endeavoring to force open the man-holes; others lying in the bottom, tightly grappled together, and the blackened faces of all presented the expression of their despair and agony."[32]

Lt. William Alexander, some forty years later, explains in some detail why the *Hunley* plunged to the bottom of Charleston Harbor in October 1863:

> The position in which the boat was found on the bottom of the river, the condition of the apparatus discovered after it was raised and pumped out, and the position of the bodies in the boat, furnished a full explanation for her loss. The boat, when found, was lying on the bottom at an angle of about 35 degrees, the bow deep in the mud. The bolting-down bolts of each cover had been removed. When the hatch covers were lifted considerable air and gas escaped. Captain Hunley's body was forward, with his head in the forward hatchway, his right hand on top of his head (he had been trying, it would seem, to raise the hatch cover). In his left hand was a candle that had never been lighted, the sea-cock on the forward end, or "Hunley's" ballast tank, was wide open, the cock-wrench not on the plug, but lying on the bottom of the boat. Mr. Park's body was found with his head in the after hatchway, his right hand above his head. He also had been trying to raise the hatch cover, but the pressure was too great. The sea-cock to his tank was properly closed, and the tank was nearly empty. The other bodies were floating in the water. Hunley and Parks were undoubtedly asphyxiated, the others drowned. The bolts that had held the iron keel ballast had been partly turned, but not sufficient to release it.
>
> In the light of these conditions, we can easily depict before our minds, and almost readily explain, what took place in the boat during the moments immediately following its submergence. Captain Hunley's practice with the boat had made him quite familiar and expert in handling her, and this familiarity produced at this time forgetfulness. It was found in practice to be easier on the crew to come to the surface by giving the pumps a few strokes and ejecting some of the water ballast, than by the momentum of the boat operating on the elevated fins. At this time the boat was under way, lighted through the dead-lights in the hatchways. He partly turned the fins to go down, but thought, no doubt, that he needed more ballast and opened his sea cock. Immediately the boat was in total darkness. He then undertook to light the candle.
>
> While trying to do this the tank quietly flooded, and under great pressure the boat sank very fast and soon overflowed, and the first intimation they would have of anything being wrong was the water rising fast, but noiselessly,

about their feet in the bottom of the boat. They tried to release the iron keel ballast, but did not turn the keys quite far enough, therefore failed. The water soon forced the air to the top of the boat and into the hatchways, where captains Hunley and Parks were found. Parks had pumped his ballast tank dry, and no doubt Captain Hunley had exhausted himself on his pump, but he had forgotten that he had not closed his sea cock.[33]

After the *Hunley* was raised, military headquarters sent the following order to Col. Alfred Rhett, commanding the Fifth Military District: "Colonel: The commanding general desires that you have the remains of Captain Hunley buried with the military honors due to an officer of his rank. He was drowned in the submarine torpedo boat some weeks since, and you will be notified by his friends when his remains are ready for interment."[34] This document clearly reinforces that Capt. Edgar Singer and some of his immediate subordinates were regarded as officers in the Confederate military. The following order regards Horace Hunley's funeral:

> Charleston, S.C., November 8, 1863. Special Order Number 19. The funeral of the late Captain Hunley will take place from the Mill's House at 4 o'clock P.M. The following officers will act as Pall Bearers, Capt. T. H. Russel Co. "A" 1st Regt. S.C. Troops, Capt. B. C. Jones Co. "B" 1st Regt. S.C. Troops, Capt. Ed Bostick Co. "E" 26th Regt. S.C. Vols., Capt. B. L. Beaty Co. "H" 26th Regt. S.C. Vols., Capt. W. W. Davis Co. "D" 26th Regt. S.C. Vols., Capt. F. H. Harleston Co. "D" 1st S.C. Arty with one 1st and one 2nd Lieutenants, and Co. "D" 1st S.C. Artillery will act as an escort. The Pall Bearers, escort and Band of 1st S.C. Artillery will assemble at the above designated place at 3 ½ o'clock P.M. precisely.[35]

Charleston military headquarters had arranged a grand funeral for Captain Hunley (the rest of the crew was buried the following day),[36] for it was specifically stated that a military band was to accompany the procession and that six "Captains" from various South Carolina regiments were to be temporarily relieved from their official duties to act as pall bearers. The *Charleston Mercury* described the grandeur that surrounded Captain Hunley's funeral:

> Monday, November 9, 1863. Last Honors to a Devoted Patriot—The remains of Captain Horace L. Hunley were yesterday interred in Magnolia Cemetery. His body was followed to the grave by a military escort, and a large number of citizens. The deceased was a native of Tennessee, but for many years past has been a resident of New Orleans. Possessed of an ample fortune, in the prime of manhood—for he was only thirty-six at the time of his death—with everything before him to make life attractive, he came to Charleston, and voluntarily joined in a patriotic enterprise which promised success, but which was attended with great peril.

Though feeling, as appears from his last letter which he wrote to his friends, a presentiment that he would perish in the adventure, he gave his whole heart, undeterred by the foreboding, to the undertaking, declaring that he would gladly sacrifice his life in the cause. That presentiment has been mournfully fulfilled. Yet who shall call that fate a sad one, which associates the name of its victim with those of his country's most unselfish martyrs?[37]

No details about the manner of Captain Hunley's death or the submarine were included, indicating the secrecy surrounding the operation. Since "a large number of citizens" escorted his body to the cemetery, many within the besieged city were most likely well aware of the circumstances and felt moved enough by his sacrifice to set aside their Sunday-afternoon activities to attend his funeral.

As Edgar Singer and some of his comrades continued operations on the eastern shore of the Mississippi River, David Bradbury was feeling the pressure from having been recently placed in charge of torpedo operations along the Texas coast.[38] The autumn of 1863 had been a difficult time for General Magruder's command because his forces were then cut off from the eastern Confederacy and hostile Indian attacks on the Texas frontier had increased to a point that soldiers from the interior, then stationed along the coast, had deserted in droves to return to their homes and protect their families.[39]

Compounding these difficulties was the continuing fear of imminent invasion from the sea, which undoubtedly placed a heavy burden on Bradbury and the numerous mechanics then fabricating torpedoes at the Port Lavaca machine shop. On November 2, 1863, a Union force landed at Brazos Santiago, Texas, and marched on Brownsville forty-eight hours later.[40] With a superior Federal force advancing toward his lightly defended city, Gen. Hamilton Bee ordered public buildings and all cotton then piled on the wharves be put to the torch. As General Bee gathered all the equipment that could be transported inland and hastily arranged the evacuation of his army, looting and indiscriminate pillaging erupted throughout the city.

General Bee's retreating army was scattered along a long front and were to regroup at King Ranch in Santa Gertrudis. From there Bee reorganized his depleted regiments, assessed what ordnance had been removed from Brownsville, and sent an urgent communication to General Magruder stating that San Antonio was not under threat of attack and that the Rio Grande region should not be evacuated. In response, Magruder immediately ordered him to proceed with his forces to the coastal city of Corpus Christi and remove as much military equipment and public property (especially the all-important government-owned cotton that could be traded for European arms and munitions) as could be transported inland.[41]

As a Federal force steadily moved unopposed up the coast from the south,

Col. W. R. Bradfute, commanding Corpus Christi, quickly sent a communication to General Magruder at Houston, informing him that he would evacuate the city and make preparations to receive the enemy at Fort Esperanza at the mouth of Matagorda Bay. Magruder immediately dispatched the 2nd and 36th Texas Cavalry Regiments southward to reinforce the isolated fort, then under the command of Col. Daniel Shea. The forces there included two batteries of his artillery as well as two or three companies of the 8th Texas Infantry and several Home Guard units.[42]

As the enemy advance was to be halted at Fort Esperanza, Captain Bradbury and other Singer operatives were ordered to reexamine the eighteen floating torpedoes they had placed in the channel between Fort Esperanza and Pass Cavallo some months earlier.[43] Bradbury filed a report two months later: "Just before the attack of Fort Esperanza, Colonel Bradfute ordered me down there with all the torpedoes we had, which were 24.... I planted them all about the trenches of the fort, with strings to set them off leading into the fort."[44] The Singer group most likely originated the plan to torpedo the channel.

They took great care in planting these devices, which "contained on an average 35 to 40 pounds of fine diamond-grained powder," and "wooden troughs or boxes, through which rawhide lines were run," had been constructed in such a way that they would not hinder the detonating process. However, the Confederates evacuated the fort following intense shelling from both land-based artillery and gunboats lying offshore. The Federal officer who took command of Fort Esperanza made the following report:

> Major: I have the honor to transmit herewith a description of the torpedoes manufactured by the "C.S.A. Torpedo Company" having a brother of Singer (of Singer's Sewing Machine notoriety) as president. I recovered and opened 13 of one kind and 3 of another; the first were applied to land defenses, and were taken from the dead angles of each salient and from the west curtain face of Fort Esperanza, at Pass Cavallo, Texas.
>
> The shape of these, was a cylinder vessel of from 13 inches to 21 inches in diameter, and from 15 inches to 24 inches in depth. These cases contained on an average 35 to 40 pounds of fine diamond-grained powder. The latter were intended for submarine defense, and were of the same pattern, with the exception of an air-vessel attached. The composition of all was tin; the latter with a prolongation of the standard, to admit of passage the striking-rod through the air vessel. To prevent against the leakage of water into the float, there was a small vertical cylinder extending between the upper and interior and the lower and interior surfaces of the float....
>
> The land torpedoes were invariably laid upon their sides, and were braced on their resistance side by a log of wood, or else stakes driven into the ground;

and to protect the machinery from clogging with sand and earth, they were covered with boards, strong enough to withstand a fair weight of earth. Around the exterior of the cases were bedded quantities of small stones, which, in the event of an explosion, would act as splinters, or makeshift projectiles, and in rainy weather would serve as drain-ways to keep the torpedo in more perfect condition than if it were fairly in contact with the earth. I am Major, very respectfully, Your Obedient Servant, J. T. Baker, ex-Captain. U.S. Engineer in charge of Fort Esperanza.[45]

This document is somewhat puzzling for it describes Singer as being the president of the "C.S.A. [Confederate States of America] Torpedo Company." No other known source of Singer-related information refers to this company name, and the reason why Baker purposely placed quotes around it is not clear. Since Captain Baker was well aware that Edgar C. Singer was the president of the rebel organization and had close family ties to the inventor of the Singer sewing machine, Baker may have received this detailed information from Confederate deserters aware of the local torpedo manufacturing facility and the background of its colorful founder.

Because Colonel Bradfute ordered the fort abandoned, the buried mines were never detonated,[46] and the first and last known use of Singer torpedoes as land mines unceremoniously came to a conclusion. Captain Bradbury concluded, "If they had made an assault on us, they (the torpedoes) would have been very destructive, but as it was we lost them."[47]

As the Texans were being forced away from the coastal regions, Captain Bradbury and his operatives quickly anchored several torpedoes at the rear of their retreating army "in a narrow channel in the lower bay, not far from the mouth of the Guadalupe River, so that the enemy cannot come up from Aransas Pass without exploding them" and "in the artificial channel between Powder Horn and Port Lavaca." Bradbury also informed his superiors in Houston that "some very powerful ones have been placed there since the evacuation of the fort. And I know that it is a dead gone thing for us to get the enemy if they make an attempt to come up; and the enemy knowing of these torpedoes is what has kept their gunboats from coming up, long before this."[48]

While Captain Bradbury and his operatives were preparing to mine the approaches of Fort Esperanza, Lieutenant Dixon (who was in Mobile trying to recruit a crew to reman the *Hunley*) received the following telegram from a greatly discouraged General Beauregard: "Charleston, S.C., November 5, 1863. Lieutenant Dixon: I can have nothing more to do with that Submarine boat, tis more dangerous to those who use it than to enemy. Sgd. G. T. Beauregard."[49]

Since Dixon and Dillingham were then under orders to go to Mobile on business concerning the *Hunley* and then return to Charleston, General Beauregard apparently had a change of heart in the weeks to come (perhaps because of the desperate conditions surrounding his command), because he later reluctantly gave his blessing to the hazardous operation. Lieutenant Dixon perhaps shared some of the general's concerns regarding the safety of the *Hunley*, because before his return to Charleston, he appointed his landlord, Henry Willey, the administrator of his estate in case he did not return. His estate was valued at "about five thousand dollars," and the contents of "one leather trunk" left in Mr. Willey's care contained many articles of clothing that only a refined gentleman of the period would have worn.[50]

From an article that appeared in the *Mobile Daily Herald* in 1904, we know that Dixon was engaged to the daughter of a Mobile steamboat captain prior to his enlistment in the Confederate army and that he was originally from Campbell County, Kentucky.[51] The article also mentions a mysterious gold coin that has become something of a Hunley legend: Before the Twenty-first Artillery Regiment left Mobile, Dixon's sweetheart handed him a twenty-dollar gold piece. The article reported that he was shot in the groin at the Battle of Shiloh by what would have been mortal wound, but the bullet was firmly embedded in the metal of the coin. From that day forward he continually carried the life-saving gold piece in his pocket. Dixon's war record notes that he was "severely wounded in the left thigh" at Shiloh.[52] The fanciful story of the gold coin turned out to be absolutely true, for the dented 1860 twenty-dollar gold piece was discovered within the hull of the *Hunley* bearing the inscription "Shiloh, April 6th 1862, My Life Preserver, G.E.D."[53]

Lieutenant Dixon and those who had accompanied him from Mobile were undoubtedly back in Charleston by the afternoon of November 12.[54] With the *Hunley* recovered in his absence, Dixon would have wanted to assess any damage sustained in either the sinking or subsequent salvage. After he learned the cause of the sinking, Dixon and his new first officer, Lt. William Alexander, likely reported their findings to military headquarters and requested permission to continue diving operations: "Charleston, S.C., November 12th, 1863. Lt. George E. Dixon Co. A. 21st Alabama Volunteers requests to be put in charge and command of the submarine torpedo boat."[55] We can only surmise what patriotic pleadings the young officer might have given to secure command of the vessel that many in Beauregard's command considered an underwater death trap, doomed to failure.

General Beauregard sent the following reply: "November 13th, 1863. To: Lieut. Dixon G. E.: The Fish Torpedo Boat having been duly transferred by me to Captain H. L. Hunley before his demise and he having cleaned and obtained it as owner, it is no longer under my control, hence I can not grant the request herein contained, but I will afford proper assistance to whoever has lawful charge of said boat and will give him authority to navigate in this bay with it, and to attack the enemy's vessels blockading this harbor. General Beauregard, Commanding."[56] The question of who was then in possession of the *Hunley* was quickly sorted out, for within twenty-four hours after Beauregard sent his letter, Thomas Jordan sent the following telegram to Dixon's superiors in Mobile: "To Col. G. Gardner, Chief of Staff, Mobile, Alabama. November 14, 1863. Please extend detail of Lieutenant George E. Dixon to end of year. (signed) Thomas Jordan, Chief of Staff Charleston, S.C."[57]

After Dixon learned he could remain in Charleston until January 1, 1864, he made the following requests:

> Charleston, November the 14th, 1863. Brig. General Jordan, Chief of Staff. Sir: Before I can proceed with my work of cleaning the Sub-Marine boat, I shall have to request of you an order on the Quartermaster or Engineer Department for ten Negroes, also an order on the Commissary Department for soap, brushes, and lime, and an order on the Arsenal to have some work done at that place. In order to make all possible haste with this work, I would be pleased to have those orders granted at your earliest convenience. I am Yours with Respect Lt. Geo. E. Dixon, Commanding Sub-Marine Boat.[58]

Within a week after burial of Captain Hunley and his crew, work was again under way to both repair and recrew the submarine for renewed operations. During this period of repair and redeployment Captain Singer, then attending meetings in Richmond, filed his first (recorded) official report, on November 13, 1863, with the Confederate Engineering Department about the various "operations" then being conducted by members of his clandestine torpedo organization.[59] The document undoubtedly discussed subjects such as early torpedo development and deployments in Texas, Louisiana, Mississippi, and Alabama (as well as Savannah, Georgia,[60] and Wilmington, North Carolina, where James Jones had been detailed on October 13);[61] the sinking of the gunboat USS *Baron DeKalb*; and the launching of the submarine in Charleston harbor.

Captain Singer was quite impressed with Dunn and Braman's designs (one of which was for a huge "ironclad torpedo boat"),[62] for shortly after Dunn's arrival in the Confederate capital, he and Singer quickly presented the plans to Senator Louis T. Wigfall from Texas (a well-known "Fire-Eater" and staunch secessionist member of the Confederate Congress, who vigorously supported both slavery and states' rights).[63] A letter from John D. Braman to his wife four months later gives us an idea of the magnitude of the plan Singer presented to the Texas senator.[64] Although Braman fails to mention anything regarding the manufacture of "*David* class" vessels (as R. W. Dunn had touched on in his letter to General Magruder),[65] much is revealed about the group's plan to build a huge iron-sheathed torpedo vessel:

We at once concentrated all the inventive genius in our party for the purpose of getting up something new that would carry destruction to the Yankees, make money for ourselves, and at the same time be of great service to the Confederacy. The result was that I got up the plan of an ironclad torpedo boat that, all who saw it admitted, was equal to the task of destroying any war ship now afloat. To carry out our plan and get our new boat under headway, it was necessary, first, to lay the whole matter before the authorities at Richmond, get their approval of the scheme, and authority, together with money and material, to build her.

Secondly, after this was accomplished the boat would have to be constructed, manned, and used. In order that the first part of our work should be properly begun, we deemed it best, after consultation, to send Dunn to Richmond, and through the influence of Wigfall and others to get the matter before Congress, and authority procured from it to carry out our plans. This part of the business Dunn was well suited for, and has succeeded in accomplishing what we desired.[66]

Saying that "I got up the plan of an ironclad torpedo boat that, all who saw it admitted, was equal to the task of destroying any war ship now afloat" might be regarded as innocent bragging to his spouse, because Dunn and Braman had first hatched the scheme after examining the *David* in Charleston.[67] Singer and Dunn expended great effort during November to convince Senator Wigfall to present the matter before Congress, which they accomplished.

Although many weeks of negotiations between the Singer group and various members of the Engineering and War Departments were yet to take place, the secretary of war ultimately approved construction of the strange vessel. Singer wrote home that "I am about to commence an enterprise that the government will watch with great interest as it promises the redemption of New Orleans and the Mississippi River. I am taking on myself a great responsibility, one that requires all my abilities. . . . I have wished a great many times that Bradbury was here, his services would be invaluable, for ingenious and energetic men are scarce."[68]

William Alexander wrote an article in *Munsey's Magazine* describing how he and Lieutenant Dixon were received at military headquarters: "We soon had the boat refitted and in good shape, reported to General Jordan that she was ready for service, and asked for a crew. After many refusals and much discussion, General Beauregard finally assented to our going aboard the Confederate receiving ship 'Indian Chief' and calling for volunteers. He strictly enjoined upon us to give a full and clear explanation of the desperately hazardous nature of the service required."[69] Although Beauregard had apparently authorized Dixon and Alexander to call for additional volunteers to recrew the submarine, surviving documents suggest that he felt compelled to clear his decision with the Navy Department: "Lieutenant Dixon is thoroughly conversant with the management of the 'Fish Boat' and maintains it offers little danger when properly handled. The destruction of the enemy's ironclads would warrant in my opinion the approval of Lt. Dixon's application (for a crew) when men are found still willing to accompany him with a full knowledge of the accidents which have already happened to the crews of that boat. Signed General Beauregard."[70]

Even though the *Hunley* had already taken the lives of thirteen men, Dixon appears to have given a speech that inspired a sufficient number of sailors to put aside their fears and step forward to recrew the submarine. According to a Confederate sailor aboard the CSS *Indian Chief* in late 1863, "Dixon had no difficulty in finding another crew of volunteers ready to take the same risks. They were Arnold Becker, C. Simkins, James A. Wicks, F. Collins and Ridgeway, all of the Confederate Navy."[71] These five men from the *Indian Chief* were joined by three volunteers who had returned with Dixon from Mobile.[72] With Dixon's crew then at full strength, the *Hunley* was once again returned to the waters of Charleston Harbor. Now that the *Hunley* was once again operational, training would have begun without delay, and new volunteers found themselves venturing into the dark depths of Charleston Harbor. Apparently, training operations went smoothly over the next few days, for on December 14 Confederate military headquarters issued the following order:

> Charleston, S.C., December 14, 1863, Special Order, Number 271. First Lieutenant George E. Dixon, Twenty-first Regiment Alabama Volunteers, will take command and direction of the Submarine Torpedo-Boat 'H. L. Hunley,' and proceed to-night to the mouth of the harbor, or as far as capacity of the vessel will allow, and will sink and destroy any vessel of the enemy with which he can come in conflict. All officers of the Confederate army in this department are commanded, and all naval officers are requested, to give such assistance to Lieutenant Dixon in the discharge of his duties as may be practicable, should he apply therefor. By command of General Beauregard.[73]

This first nocturnal patrol that Dixon and his untested crew attempted was undoubtedly an awkward experience plagued by mistakes and uncertainty. According to William Alexander, the strong currents encountered at the entrance of the harbor were something of a surprise, which caused immediate problems with the long tow line that dragged the torpedo in their wake: "The torpedo was a copper cylinder holding a charge of ninety pounds of explosive, with percussion and primer mechanism, set off by triggers. It was originally intended to float the torpedo on the surface of the water, towed by the boat, which was to dive under the vessel to be attacked. In experiments made with some old flat boats in smooth water, this plan operated successfully, but in a seaway the torpedo was continually coming too near our craft."[74]

During the course of the long night of struggling against strong currents, Dixon quickly discovered that the prolonged cranking from their dock at Mount Pleasant had caused unnecessary fatigue on his crew. Following this first attempt, the steam-powered *David* towed the submarine past Fort Sumter whenever weather conditions allowed. The *David*'s skipper, Chief Engineer James Tomb, wrote about the orders given him: "General Beauregard placed Lieutenant Dixon in charge of her [the submarine] and requested of flag officer Tucker, that the David tow the Hunley past Fort Sumter when he concluded to run out over the bar, so that the crew would be fresh to work her."[75]

According to Tomb's and Alexander's recollections, mid- to late December was a busy time for both the *David* and crew of the *Hunley*: "During this time we went out on an average of four nights a week."[76] Within forty-eight hours after Lieutenant Dixon had received orders to "proceed to-night to the mouth of the harbor" and "destroy any vessel of the enemy with which he might come in conflict," the Confederate Engineering Department sent a curious letter to Gen. W. H. Whiting (military commander of Wilmington, North Carolina). The general had apparently recently contacted the secretary of war about torpedo boats and inquired if the services of Singer operative Robert W. Dunn could be secured to oversee construction of such boats in Wilmington. The Engineering Department's response never discussed why the general regarded Dunn as an expert on the construction of such vessels or how his talents had been brought to the general's attention:

> December 16, 1863, Major General W. H. Whiting, Headquarters, Wilmington, N.C. General: Your letter of the 30th requesting that plans of torpedo boats be sent you has been referred to this Bureau for reply by the Honorable Secretary of War. On conference with Captain James, Chief Engineer for your department, I find that the Bureau has no drawings on that subject different from those already in your possession. Mr. Dunn, alluded to in your letter will be informed of your wish that he should proceed to Wilmington to aid in the construction of the torpedo boats.[77]

The following inquiry was delivered to Captain Singer's Richmond headquarters the same day: "December 16, 1863, R. W. Dunn, Spotswood Hotel, Richmond. Sir: In a letter recently received from Maj. Gen W. H. Whiting, Commanding at Wilmington N.C. he inquires whether you are an expert in the construction of torpedo boats, and says that, in such case, you could render valuable service in his Department. Please communicate with this Bureau on the subject."[78]

As noted previously, for well over a month both he and Captain Singer had been in Richmond meeting with Senator Louis Wigfall, attempting to bring their idea of fabricating large ironclad torpedo boats before the Confederate Congress.⁷⁹ With these events established, combined with the fact that Dunn was a part owner of the *Hunley*, his name probably had casually been circulated throughout the Richmond Engineering Department as someone familiar with torpedo boat construction. General Whiting may have learned of Dunn's knowledge of the subject from James Jones, who had been ordered to Wilmington on October 13, but documents are not available to confirm this.⁸⁰

The Confederate Engineering Department also contacted Captain Singer concerning the general's request: "Richmond, Va. December 23, 1863, E. C. Singer and Company, Care of Col. D. Harris, Chief Engineer, Charleston, S.C. Gentlemen: Please furnish this bureau as promptly as possible with the plan of a torpedo boat, best adapted in your opinion for the purposes of harbor defense."⁸¹ This communication is extremely important to the history of the *Hunley* because it places "E. C. Singer and Company" back in Charleston at exactly the same time that Dixon and his volunteer crew renewed operations (although it does not mention how long they had been in the city). Since James McClintock had also returned to Charleston (if indeed he had ever left),⁸² the Singer group was most likely then focusing their attention on making their South Carolina submarine venture a success (they may also have been fabricating Singer-Fretwell torpedoes for the Charleston defenses).

We can assume that some communication between Captain Singer, Robert Dunn, and General Whiting transpired, for at least two "*David* class" torpedo boats and another of a much larger dimension were constructed at the Wilmington Navy Yard in the months that followed (perhaps under the direction of James Jones, who remained on duty in Wilmington until the late fall of 1864).⁸³ Singer-Fretwell torpedo strikes that were reported to have taken place during this time on the York River near Richmond are not substantiated. In March 1864, John D. Braman wrote a letter to his wife in Texas that outlined several projects that the Singer organization was about to undertake or had recently finished. After discussing the group's plan to construct a large "ironclad torpedo boat," Braman penned the following brief line about recent Singer-Fretwell torpedo strikes in Virginia: "Besides this, we sunk one boat and seriously injured another in the York River in December, the full particulars of which Dunn will tell you when he sees you."⁸⁴

As with Robert Dunn's letter to General Magruder concerning the sinking of the Union troop transport *Gray Cloud*, Braman's alleged December 1863 destruction of a Federal vessel in the York River cannot be verified from documentation found in Union naval records.⁸⁵ Perhaps the incident was nothing more than the unrecorded sinking of a small Federal steam launch, manned by Union pickets patrolling the river. Dunn wrote to General Magruder in April 1864 about what he knew of the alleged encounter: "A gunboat in York River (name not recalled) was blown up by one of them [our torpedoes], destroying the boat and crew and injuring another boat badly that lay near by."[86] Since Robert Dunn was attending meetings in Richmond during the period in question, we can conclude that he had firsthand knowledge of the encounter, because he may have been working with the Singer operatives responsible for anchoring the devices. Without Union corroboration, we are left with nothing more than Dunn and Braman's puzzling claims.

While Robert Dunn continued his duties at the Confederate capital, and presumably communicated with General Whiting regarding construction of torpedo boats in Wilmington, Dixon continued to go to sea whenever weather conditions allowed. Headquarters looked favorably on the secret nocturnal ventures into enemy-controlled waters since just five days into the new year they sent the following request to Mobile: "January 5, 1864. To Colonel Gardner Chief of Staff. Please extend Lieutenant Dixon's detail 30 days longer."[87]

On the night of January 5, two Confederate seamen named Shipp and Belton surrendered to Federal pickets and were immediately interrogated about events then taking place in Charleston. After being questioned about harbor defenses and the capabilities of the *David* torpedo boat, one of the men discussed the *Hunley* in great detail, giving the following testimony:

> She is about 35 feet long, height about the same as the "David" (5 ½ feet). Has a propeller at the end, she is not driven by steam, but her propeller is turned by hand. Has two man heads on the upper side about 12 or 14 feet apart. the entrance into her is through these man heads, the heads being turned back, they are all used to look out of.
>
> She has had bad accidents hitherto—but was owing to those in her not understanding her. Thinks that she can be worked perfectly safe by persons who understand her. Can be driven 5 knots an hour without exertion to the men working her. When she went down the last time, was on the bottom 2 weeks before she was raised, manheads are about 16 inches high, and are just above water when trimmed. Believes she was brought here about 1st September. . . . When she went down the last time, was on the bottom 2 weeks before she was raised. Saw her when she was raised the last time. They then hoisted her out of the harbor, refitted her and got another crew. Saw her after that submerged, Saw her go under the Indian Chief, and then saw her go back under again. She made about ½ mile in the dives.
>
> Saw her dive under the Charleston—went under about 250 feet from her and come up about 300 feet beyond her, was about 20 minutes under the water. . . . Believes she is at Mount Pleasant. One of her crew who belongs to his vessel came back for his clothes and said she was going down there as a station.[88]

Within hours after this testimony was recorded, Admiral Dahlgren issued the following orders to the fleet:

> Order No. 2, January 7, 1864. I have reliable information that the Rebels have two torpedo boats ready for service, which may be expected on the first night when the weather is suitable for their movement. One of these is the "David" which attacked the "Ironsides" in October, the other is similar to it.
>
> There is also one of another kind, which is nearly submerged, and can be entirely so; it is intended to go under the bottoms of vessels and there operate. This is believed by my informant to be sure of well working, though from bad management it has hereto met with accidents, and was lying off Mount Pleasant two nights since. There being every reason to expect a visit from some or all of these torpedoes, the greatest vigilance will be needed to guard against them.
>
> The ironclads must have their fenders rigged out, and their own boats in motion about them. A netting must also be dropped overboard from the ends of the fenders, kept down with shot, and extended along the whole length of the sides; howitzers loaded with canister on the decks and a calcium [search light] for each monitor. The tugs and picket boats must be incessantly upon the lookout, when the water is not rough, whether the weather is clear or rainy. I observe the ironclads are not anchored so as to be entirely clear of each other's fire if opened suddenly in the dark. This must be corrected, and Captain Rowan will assign the monitors suitable positions for this purpose, particularly with reference to his own vessel.
>
> It is also advisable not to anchor in the deepest part of the channel; for by not leaving much space between the bottom of the vessel and the bottom of the channel, it will be impossible for the diving torpedo to operate except on the sides, and there will be less difficulty in raising a vessel if sunk.[89]

Admiral Dahlgren learned of the *Hunley*'s nocturnal operations at about the same time that they were about to be temporarily suspended, for within seventy-two hours after this order was issued, both Lieutenants Dixon and Alexander would be searching for another location to moor the *Hunley* on nearby Sullivan's Island. Engineer Tomb explains that the operation was temporarily suspended because towing the submarine was more dangerous than anyone had suspected: "The last night the 'David' towed him [Dixon] down the harbor, his torpedo got foul of us and came near blowing up both boats before we got it clear of the bottom, where it had drifted. I let him go after passing Fort Sumter, and on my making report of this, Flag-Officer Tucker refused to have the 'David' tow him again."[90]

Since Flag Officer Tucker abruptly canceled the towing services from the *David* (probably on either January 8 or 9), Dixon and Alexander had to find another mooring from which to strike out at the blockading fleet.

Alexander explains: "On account of chain booms having been put around the 'Ironsides' and the monitors in Charleston harbor to keep us off these vessels, we had to turn our attention to the fleet outside."[91] On the morning of January 10, 1864, Lieutenants Dixon and Alexander crossed the narrow wooden bridge that connected Mount Pleasant and Sullivan's Island to seek out a new mooring closer to the open sea. Folded in Lieutenant Dixon's pocket was the following order from Charleston headquarters to Brig. General W. S. Walker, commanding the Third Military District: "Charleston, S.C. January 10th, 1864. General: I am instructed by the commanding General to inform you that Lieutenant Dixon, the bearer of this, goes to your district for the purpose of monitoring the several positions of the enemy in your front with a view to operations (by water) against the enemy. The commanding General directs that you furnish Lt. Dixon with every possible facility for carrying out his plans."[92]

The two engineering officers were satisfied that a sand fortification named Battery Marshall, located on Breach Inlet at the northern tip of the island, could be utilized as a successful base of operations: "We were ordered to moor the boat off Battery Marshall, on Sullivan's Island. The nearest vessel which we understood to be the United States Frigate Wabash, was about twelve miles off, and she was our objective point from this time on."[93]

At the same time that Lieutenants Dixon and Alexander were preparing to move their submarine to the new base of operations, David Bradbury was sending a report concerning the activities that had taken place within Texas since the formation of Captain Singer's torpedo company:

Victoria, January 11, 1864. Capt. E. P. Turner, A. A. G. Captain: I have the honor to enclose communication from Captain Bradbury, in relation to his torpedoes. I have the honor etc. James Duff, Colonel, Commanding etc.

Victoria, January 9, 1864. Col. James Duff, Commanding. Colonel: I came up from Port Lavaca, in compliance to your request, and have this to report in relation to the Singer torpedo: In the month of May last, and by order of Major Shea, I placed 18 floating torpedoes in the channel between Fort Esperanza and the bar at Pass Cavallo, but the channel was very wide, water deep (30 feet) and current strong, and I made up my mind when I put them there that it was doubtful if they ever did any good, and the length of time that they have been in makes them useless now.

But I have good reason to believe that the fact of their being there, and they knowing it, has kept them out till they got here in force. Just before the attack on Fort Esperanza, Colonel Bradford ordered me down there with all the torpedoes we had here, which was 24. I planted them all about the trenches of the fort, with strings to set them off leading into the fort, and if they had made an assault on us they would have been very destructive, but as it was we lost them.

We have 4 fresh ones in a narrow channel in the lower bay, not far from the mouth of the Guadalupe River, so that the enemy cannot come up from Aransas Pass without exploding them. In the artificial channel between Powder Horn and Port Lavaca we have quite a large number of them. Some are a year old, but some very powerful ones have been placed there since the evacuation of the fort, say within five weeks. And I know that it is a dead gone thing for us to get the enemy if they make an attempt to come up; and the enemy knowing of these torpedoes is what has kept their gunboats from coming up, long before this.

I regret that I am not able to report my positive destruction of the enemy's property by the torpedoes up to this time, but if they will not try them we cannot expect to destroy them; but I think it ought to be put down to our credit that we have kept them out of Pass Cavallo for one year, and that we are still keeping them out of Lavaca Bay.

We have plenty more torpedoes here at Victoria, and after the lapse of three of four weeks more I would give it as my opinion that some more fresh ones should be put in the channel between Powder Horn and Lavaca, but at the present moment it certainly does not need any more. In regard to placing one in the bay between Indianola and Matagorda it cannot be done to any advantage, and for sundry reasons: First because the bay is so very wide that it would take a large number to give the least probability of success; next, boats and boatmen are not to be had, and if they were the danger of being picked up by the enemy is very great.

I am very anxious to have a success with the torpedoes, and beg to assure General Magruder, through you, that I am constantly on the lookout for an opportunity to use them, and shall let no opportunity pass without using my best endeavor to improve it. On the other side of the Mississippi River we have had some splendid success with them, and I hope to yet on this side before the war is over, and if I don't it shall not be for want of faithful trying on my part. Respectfully, D. Bradbury.[94]

Captain Bradbury was obviously under the impression that Edgar Singer and his eastern operatives had by early January 1863 achieved "some splendid success" with their torpedoes on the eastern side of the Mississippi River. Although only the July sinking of the Federal gunboat *Baron DeKalb* can at present be verified, Bradbury felt very optimistic that similar torpedo successes would soon take place in Texas waters.

While David Bradbury oversaw torpedo operations and studied maps of the Texas coast with his superiors for the best locations to anchor mines, Adm. John Dahlgren was sending an urgent communication to the secretary of the navy informing him about the new underwater threat to his fleet. After reacquainting the secretary with the design of the *David* torpedo boat that had unsuccessfully attacked the USS *Ironsides* in early October, Dahlgren wrote about the *Hunley*:

> The "Diver" as she is called is also ready, and with the original "David" is now at Mount Pleasant on the lookout for a chance. The action of the "David" has been of course pretty well exemplified on the "Ironsides"—That of the "Diver" is different, as it is intended to submerge completely, get under the bottom, attach the torpedo, haul off and pull trigger. So far the trials have been unlucky, having drowned three crews of seventeen men in all. [The admiral was misinformed as to the number of accidental sinkings.] Still she does dive, as one of the deserters saw her pass twice under the bottom of the vessel he was in, and once under the Charleston.[95]

While Dixon and Alexander made preparations to redeploy the *Hunley* from the harbor to Breach Inlet, the recently returned Captain Singer was preparing to redesign the means by which his submarine would deliver its explosive device. Chief Engineer James Tomb describes the new torpedo configuration: "Lieutenant Dixon and myself discussed the best plan to make use of the torpedo boat and decided that the adjustable spar used on the David was the best, and he said he would use it on the Hunley, keeping her on the surface and the torpedo some eight feet below."[96]

Instead of towing a contact torpedo at the end of a long rope miles out to sea in unpredictable currents, the *Hunley* would now attack the wooden ships in the outer circle of the blockade with an explosive charge attached to a spar assembly extending from the bow. An 1870 article in the *Charleston Daily Republican* discusses General Beauregard's involvement in this decision: "General Beauregard changed the arrangement of the torpedo by fastening it to the bow. Its front was terminated by a sharp and barbed lance-head so that when the boat was driven end on against a ship's sides, the lance head would be forced deep into the timbers below the water line, and would fasten the torpedo firmly against the ship. Then the torpedo boat would back off and explode it by a lanyard."[97]

To further reinforce General Beauregard's involvement in the decision process that brought about the reconfiguration of the *Hunley*'s explosive device, Capt. Francis Lee, designer of the spar arrangement adopted by Captain Singer and his torpedo engineers in early January 1864, sent General Beauregard a letter, discussing how he had invented his spar assembly in the summer of 1861 and reminded the general:

> You were in command of the department when the attack was made upon the Housatonic with the "Fish Boat," or at least you were when the attack was determined upon. Lieutenant Dixon proposed using his craft as she was originally designed to be used, i.e. as a diving boat, carrying a drift torpedo,

fired by triggers. But considering the brave lives already sacrificed on board her you refused to permit him to go unless he would consent to use the spar torpedo. I opposed, as far as it was in my power, to use my torpedo, as the craft had little buoyancy and the probability of further disaster seemed to be almost certain. Dixon, however, insisted and carried his point, and I superintended in person the placing of a 50 pound torpedo on a spar at the bow of his craft. . . . The conversation above alluded to, occurred at your headquarters, you being present together with Dixon, myself and I think General Jordan.[98]

Singer and various members of his Secret Service Company may well have been present at this meeting, since Lieutenant Dixon probably would not have had the authority to change the submarine's torpedo configuration without Singer's consent. A spar configuration was quickly adapted to the *Hunley* shortly after Flag Officer Tucker suspended the *David*'s towing of the vessel. The explosive device attached to the end that would deliver the intended death blow was entirely the invention of Capt. Edgar Singer and his torpedo engineers.[99] An article in the *U.S. Naval Institute Proceedings* describes how the torpedo was to be detonated:

A torpedo was designed which could be mounted on a short pole and which would delay its explosion until the attacking vessel could back off to a safe position. It consisted of a steel head which fitted as a thimble over the end of the ten foot spar or pipe projecting from the bow. This was driven into the enemy's wooden hull by ramming and was retained there by saw-toothed corrugations when the fish boat backed off. As it slipped off the spar, it would keep with it the torpedo, which was a simple copper can of powder fitted with a trigger. This trigger was attached to a cord lanyard carried on a reel on deck and after the boat had backed a safe distance, 150 yards, the rope was to tighten and would trip the trigger.[100]

Even after the *Hunley* had been relocated to Breach Inlet, the crew continued to be quartered at Mount Pleasant several miles away. During this period, the crew of the *Hunley* presumably practiced this new method of attack behind Battery Marshall in an area known as the back bay. Alexander gives us a glimpse into the crew's daily activities during this period of redeployment: "Our daily routine, whenever possible, was about as follows: Leave Mount Pleasant about 1 P.M., walk seven miles to Battery Marshall on the beach (this exposed us to fire, but it was the best walking), take the boat out and practice the crew for two hours in the Back Bay."[101] The practice sessions probably consisted of simulated attacks with their new torpedo assembly against an old vessel moored behind Battery Marshall. By January 19, Dixon and Alexander had apparently become confident that Captain Singer's new torpedo configuration could be used effectively, because Confederate military headquarters issued these orders:

> First Lieutenant George E. Dixon, 21st. Regiment Alabama Volunteers, will take command and direction of the Submarine Torpedo Boat—"H. L. Hunley"—and proceed at his discretion to the mouth of the harbor, or as far as the capacity of his vessel will allow, and will sink and destroy any vessel of the enemy with which he can come in conflict. All officers of the Confederate Army in this Department are commanded, and all naval officers are requested, to give such assistance to Lt. Dixon in the discharge of his duties as may be practicable, should he apply therefore. By order of General Beauregard.[102]

While rumors of the "Diving Torpedo" spread throughout the Union fleet in the weeks following Admiral Dahlgren's January 7 warning, Dixon, Alexander, and their crew continued to practice. Alexander describes how the submarine was to attack the wooden-hulled ships in the outer circle of the blockade: "The plan was to take the bearings of the ships as they took position for the night, steer for one of them, keeping about six feet under water, coming occasionally to the surface for air and observation, and when nearing the vessel, come to the surface for final observation before striking her, which was to be done under the counter, if possible."[103]

Within forty-eight hours after Lieutenant Dixon had received his revised orders, Admiral Dahlgren was examining a wooden model of the submarine that had been fabricated by one of the deserters from the CSS *Indian Chief*. After studying the design with several officers in his command, the admiral forwarded the model to the secretary of the navy with the following description:

> January 22, 1864. The Honorable Gideon Welles, Secretary of the Navy Washington, D.C. Sir: I transmit by the "Massachusetts," a model of the "Diver" which is said to have been built at Mobile by the rebels, and brought to this place for use against the vessels in this Squadron. The Department will find a brief description of her in my communication of the 13th. The model was made by E. C. Belton who is a mechanic, and ran an engine on the Montgomery and Mobile Railroad for some time. He worked in a building near where it was built and claims to understand fully its construction. It has been very unlucky in the trials made with it, and is stated to have drowned at different times three crews of seventeen men.[104]

While Admiral Dahlgren kept Washington informed about the new underwater threat to his fleet, Lieutenants Dixon and Alexander, unaware that the *Hunley*'s existence had been compromised to the Federals, prepared their submarine for another nocturnal attempt on the wooden ships far out to sea.

Alexander describes the late-afternoon ritual that he and Dixon observed as the sun slowly settled over the cold winter marshes behind Battery Marshall: "Dixon and myself would then stretch out on the beach with the compass between us and get the bearings of the nearest vessel as she took her position for the night; ship up the torpedo on the boom, and when dark, go out, steering for that vessel, proceed until the condition of the men, sea, tide, wind, moon and daylight compelled our return to the dock; unship the torpedo, put it under guard at Battery Marshall, walk back to quarters at Mount Pleasant and cook breakfast."[105]

The physical drain on the crew of the *Hunley* during this period must have been tremendous, for after making the seven-mile trek from their quarters at Mount Pleasant, Alexander informs us that they would "take the boat out and practice the crew for two hours in the Back Bay" before going to sea until the following dawn.[106] What sleep the men may have gotten during daylight hours is uncertain, because Charleston was under a slow bombardment from recently established Union batteries placed at the northern tip of Morris Island during this period. Alexander recollects the many dangers that faced Dixon's crew:

> It was Winter, therefore necessary that we go out with the ebb and come in with the flood tide, a fair wind and a dark moon. This latter was essential to success, as our experience had fully demonstrated the necessity of occasionally coming to the surface, slightly lifting the after hatch-cover and letting in a little air. On several occasions we came to the surface for air, opened the cover and heard the men in the Federal picket boats talking and singing. During this time we went out on an average of four nights a week, but on account of the weather, and considering the physical condition of the men to propel the boat back again, often, after going out six or seven miles, we would have to return. This we always found a task, and many times it taxed our utmost exertions to keep from drifting out to sea, daylight often breaking while we were yet in range.[107]

As a result of their surprise nocturnal encounters with Federal picket boats and the unavoidable predicament of occasionally being caught at sea after sunrise, Dixon and Alexander decided that a submerged endurance test should be conducted to establish the length of time their crew could remain underwater without exhausting the air supply. Alexander explains in graphic detail how he and Dixon conducted this dangerous experiment:

> This experience, also our desire to know, in case we struck a vessel (circumstances required our keeping below the surface), suggested that while in safe water we make the experiment to find out how long it was possible to stay under water without coming to the surface for air and not injure the crew. It was agreed to by all hands to sink and let the boat rest on the bottom, in the Back

bay, off Battery Marshall, each man to make equal physical exertion in turning the propeller. It was also agreed that if anyone in the boat felt that if he must come to the surface for air, and he gave the word "up," we would at once bring the boat to the surface. It was usual, when practicing in the bay, that the banks would be lined with soldiers. One evening, after alternately diving and rising many times, Dixon and myself and several of the crew compared watches, noted the time and sank for the test. In twenty-five minutes after I had closed the after manhead and excluded the outer air the candle would not burn. Dixon forward and myself aft, turning on the propeller cranks as hard as we could.

In comparing our individual experience afterwards, the experience of one was found to have been the experience of all. Each man had determined that he would not be the first to say "up!" Not a word was said, except the occasional, "How is it," between Dixon and myself, until it was as the voice of one man, the word "up" came from all nine. We started the pumps. Dixon's worked all right, but I soon realized that my pump was not throwing. From experience I guessed the cause of the failure, took off the cap of the pump, lifted the valve, and drew out some seaweed that had choked it.

During the time it took to do this the boat was considerably by the stern. Thick darkness prevailed. All hands had already endured what they thought was the utmost limit. Some of the crew almost lost control of themselves. It was a terrible few minutes, "better imagined than described." We soon had the boat to the surface and the manhead opened. Fresh air! What an experience! Well, the sun was shining when we went down, the beach lined with soldiers. It was now quite dark, with one solitary soldier gazing on the spot where he had seen the boat before going down the last time. He did not see the boat until he saw me standing on the hatch combing, calling to him to stand by to take the line. A light was struck and the time taken. We had been on the bottom two hours and thirty-five minutes. The candles ceased to burn in twenty-five minutes after we went down, showing that we had remained under water two hours and ten minutes after the candle went out.

The soldier informed us that we had been given up for lost, that a message had been sent to General Beauregard at Charleston that the torpedo boat had been lost that evening off Battery Marshall with all hands. We got back to our quarters at Mount Pleasant that night, went over early next morning to the city (Charleston), and reported to General Beauregard the facts of the affair. They were all glad to see us.

After making a full report of our experience, General Rains, of General Beauregard's staff, who was present, expressed some doubt of our having stayed under water two hours and ten minutes after the candle went out. Not that any of us wanted to go through the same experience again, but we did our best to get him to come over to Sullivan's Island and witness a demonstration of the fact, but with-out avail.[108]

It is extremely fortunate that Lieutenant Alexander's writings regarding his adventures aboard the *Hunley* were published the *New Orleans Picayune* and *Munsey's Magazine,* for without them history would be completely silent about the activities of the submarine while based at Battery Marshall. Judging from the lack of information found in the Charleston order books, after the January 20 directive that ordered Dixon to take the submarine to sea at his "discretion," nothing more is recorded on the subject of the *Hunley* or the crew (other than a couple of insignificant requests).

Alexander described what it was like to venture into enemy-controlled waters, night after night, in the iron hull of a damp submarine that had already claimed the lives of thirteen men:

> We continued to go out as often as the weather permitted, hoping against hope, each time taking greater risks of getting back. On the last day of January we interviewed the Charleston pilots again, and they gave it as their opinion that the wind would hold in the same quarter for several weeks. . . .[109]

> We continued to go out . . . covering a longer distance when the wind was off shore, until at last we demonstrated to our satisfaction that we could reach the blockading squadron, as we could cover over twelve miles when the sea was comparatively quiet. After notifying General Beauregard of the success of our experiments, it was decided to make an attack on the first clear night when a land breeze was blowing. . . . I think that all felt as I did at the time. We had proved that the craft could be successfully operated both above and beneath the surface, in spite of the many fatalities which she had caused, and I don't believe a man considered the danger that awaited him. The honor of being the first to engage the enemy in this novel way overshadowed all else.[110]

As Dixon's thirty-day extended duty once more came to an end, General Jordan sent the following message to Alabama: "To: Col. Garner, Chief of Staff Mobile. February 5, 1864. Must again ask extension of detail of Lieutenant George E. Dixon for 30 days. Signed General Jordan."[111] It seems Dixon's unique services were too valuable to permit his return to his post in Alabama. The request was again granted from General Maury's headquarters.

By a strange twist of fate, at precisely the same time that Dixon was having his detached duty in Charleston extended, an order from Mobile was being delivered to Charleston headquarters that would injure the morale of the *Hunley*'s crew and jeopardize the submarine's continued efficiency: According to Alexander,

> On February 5, 1864, I received orders to report in Charleston to General Jordan, chief of staff, who gave me transportation and orders to report at Mobile, to build a breech-loading, repeating gun (cannon). This was a terrible blow, both to Dixon and myself, after we had gone through so much together.

General Jordan told Dixon he would get two men to take my place from the German artillery, but that I was wanted in Mobile.

It was thought best not to tell the crew that I was to leave them. I left Charleston that night and reached Mobile in due course. I received from Dixon two notes shortly after reaching Mobile, one stating that the wind still held in the same quarter, etc., the other telling the regrets of the crew at my leaving and their feelings towards me; also that he expected to get men from the artillery to take my place. These notes, together with my passes are before me as I write. What mingled reminiscences they bring![112]

Alexander states directly that the number of men to be selected from the "German artillery" after his departure was two. Alexander might have been mistaken about the actual number of crewmen called away from duty in early February, for he wrote this article almost forty years after the event had taken place. In his *Picayune* article, he touched on this puzzling discrepancy without providing any clarification; however, a line in his *Munsey Magazine* article, written the following year, may clarify why two new volunteers were needed: "All of the crew who had toiled and risked death during those long and weary months were in their places except myself and one other, also ordered to special duty."[113]

The "one other, also ordered to special duty" was undoubtedly one of the two men who had accompanied Dixon and Alexander from Mobile. And since we know that a man named Miller, whom Alexander mentions in an 1899 letter as having gone to South Carolina with him and Dixon,[114] is known to have stayed in Charleston, the mystery crewman reassigned to "special duty" most likely was Henry Dillingham, who had been ordered with Dixon to go to Mobile on business concerning the submarine.[115] The original February 5 order has not yet come to light, leaving the identity of this unknown individual open to speculation.

From all surviving sources of *Hunley*-related information, we can conclude that Singer returned to the Confederate capital soon after designing and fabricating the new torpedo that would be utilized with a spar assembly. Congressional leaders and Engineering Department officers probably requested his presence, since contract negotiations regarding the fabrication of large ironclad torpedo boats were drawing to a close. An interview in 1916 notes that in late January "Captain Singer then prepared a torpedo and attached it to the boat, after which he went to Richmond, where he remained while waiting results from the Hunley and her daring crew."[116]

In Richmond, Captain Singer and Dunn reached an agreement with the Confederate War Department regarding construction of ironclad torpedo boats. Within seventy-two hours after Alexander had received his orders to report to Mobile, Dunn was preparing to journey back to Texas (by way of

Mobile) to personally submit plans to Gen. Kirby Smith for the proposed construction of such vessels for his Trans-Mississippi Department.[117] While Dunn made arrangements to return to Mobile, before attempting the hazardous crossing of the Mississippi River under cover of darkness, Singer sat in his room at the Spotswood Hotel and penned a lengthy letter to his wife in Texas. Some weeks later while the letter was in transit through enemy-controlled territory, it, along with many others, fell into Federal hands.

Upon reading the captured communications, Union intelligence officers stumbled across Captain Singer's informative letter, immediately penned a copy, and along with several others that contained militarily significant information, forwarded it to Washington. From the captured pages, now on file at the National Archives (bearing the words "Copies sent to Secretary of War" on the upper corner), comes one of only two known wartime letters written by Edgar C. Singer:

> Richmond. February 8, 1864. Dear Wife: I wrote you yesterday a short letter by a Mr. Adams that lives at Harrisburg, six miles below Houston, Mr. Dunn leaves for Texas tomorrow. It is with my greatest reluctance that I remain, but after mature judgement I think it my duty not only to the country but to you and the children. I assure you if I thought Bradbury would neglect you I would return, I have nothing else to work for.
>
> I have had a long talk with Dunn and have told him what he and Bradbury must do, that under no consideration whatever, are you to be neglected, you must have what money you want so that if you want to move from Texana to Lavaca, or where you please, you can do it. I am certain the sacrifice that I make in remaining here entitles you to special attention, we will get all the assistance from the government we want and have no doubt of our success, in fact we have presented the only feasible plan to the government which they all agree to it if success attends us, the result will be great, you must not be backward about asking them for money. It is their duty to give it to you because they will be reimbursed and share the profits with me, they may think that I am over particular, but I am not, and if there is any failure in my request I shall quit and return home. My mind must not be troubled with matters of this kind.
>
> I am about to commence an enterprise that the government will watch with great interest as it promises the redemption of New Orleans and the Mississippi River. I am taking on myself a great responsibility, one that requires all my abilities.... Everybody has gone mad with speculation and patriotism is at an end, thousands of women and children are in a state of destitution, their protectors having gone to the army with the understanding that those left at home were to clothe and feed their families during their absence, which

understanding has long since been forgotten. My opportunities have been very good to see the state of things as they really are, having traveled over the different states several times that are now in our possession. The people are badly demoralized, the soldiers in the army will be unfit for civil life, at least a large portion of what are left of this awful and bloody tragedy.

The contest this spring will be a bloody one, the government is placing all in the army as though the crisis were at hand which I am satisfied of myself, I believe we will win, but at what a fearful cost. . . .

The army as a general thing is in very good spirits, but the people are very desponding, however a little success would soon regulate that. The state of our currency is a very perplexing question, as yet there have been no laws passed in regard to it yet, all are watching with great anxiety. My proposition is to tax the property heavily and give the federals a good whipping and the money will be good and not until then. I have but little news to write you this morning, we got news of the federals making a raid in this direction, they are within 17 miles of Richmond, but not in sufficient force to reach Richmond.

Willcox, congressman from San Antonio died this morning of apoplexy, the city is filled with great men, there has been as high as 6 or 8 generals stopping at the Spotswood House where we are at all times, besides innumerable quantities of colonels, captains, lieutenants etc. General John Morgan was here a short time ago and lionized to a great extent. He is a modest, fine looking man and deserves the gratitude of the nation.

I spend a portion of my time either in the upper or lower houses of congress to hear the discussions upon different bills that come up before them, it being a scene that I never witnessed before renders it very interesting to me, still it does not come up to my expectation of a body of great men. I had a letter from Henry Bradbury a short time ago stating the boys were all well and in good spirits, but very anxious to go to Texas. I am going to manage some way to see them, they are now at Cave Springs Georgia, not far from Rome. I wish to god they could get sent to the other side of the river, their old captain has been ordered to report on the other side and I would not be surprised if the boys did not take a notion to follow him. I have wished a great many times that Bradbury was here, his services would be invaluable for ingenious and energetic men are scarce. E. C. Singer.[118]

At this time, Lieutenant Dixon and the crew of the submarine boat were in all likelihood welcoming the services of Corp. C. F. Carlson of Company A, South Carolina Light Artillery, who had recently volunteered for service aboard the *Hunley*. Soon after Carlson arrived, a new federal sloop-of-war began to be seen dropping evening anchor in the vicinity of Rattlesnake Shoal, just over three miles from the mouth of Breach Inlet. With orders to keep up a full head of steam throughout the night and run down or destroy any blockade runner that attempted to pass, Capt. Charles Pickering and the crew of the 1,240-ton, steam sloop-of-war *Housatonic* rocked at anchor within sight of Breach Inlet, oblivious to the fact that the "diving torpedo" lay quietly moored not fifty yards beyond its entrance.

With a menacing sloop-of-war anchored offshore, Dixon chose to change tactics in favor of a bold new plan of attack. Since they could approach the vessel within a couple of hours after nightfall, he decided that the *Hunley* would attack the sloop on the first calm evening, and once clear of its intended victim, signal the watchtower at Battery Marshall to light a beacon fire at the mouth of Breach Inlet. Dixon and his crew would then steer for the light before the expected steam-powered federal picket boats could converge on the area. This attempt may well have been the first time that an arranged signal would be utilized to alert the shore battery to expose a homing beacon for Dixon's crew to follow.[119]

While Lieutenant Dixon finalized his plan of attack, the sailors on the USS *Housatonic*, anchored some miles offshore, wallowed in heavy seas, night after night, guarding Rattlesnake Shoal against any rebel ship attempting to breach the inner circle of the Federal blockade. To guard against Confederate torpedo vessels that might approach undetected during the night, Captain Pickering posted six well-armed lookouts on the *Housatonic*'s deck, one on each gangway, cathead, and quarter. The officer of the deck stood on the bridge, with two other officers posted at the quarter deck and forecastle, each scanning the dark horizon with telescopes, hoping to discover a blockade runner.

At the anchor chains, two sailors stood at the ready to slip the chain and set the *Housatonic* free at the first sign of trouble. In the smoky engine room below decks, soot-covered sailors continually shoveled coal on the hot boiler fires throughout the night, ensuring that the warship could steam forward at a moment's notice. On the cold windswept decks above, shivering sailors sang songs and played cards around their loaded cannons, ready at a moment's notice to fire into any rebel blockade runner that came within range.[120]

From the shore near the mouth of Breach Inlet Lieutenant Dixon and his well-practiced crew could see the faint lights of this powerful new sloop as it rocked at anchorage, all of them hoping that the sloop would remain at its new station until the rough seas and winds subsided long enough for them to attack. This winter weather system severely curtailed the *Hunley*'s movements through much of early and mid-February. The *Charleston Daily Courier* reported on February 17, 1864, that the offshore winds were so severe at this time that several Federal ironclads were observed at anchor in Light House Inlet (on Morris Island) after dark to escape the heavy seas.

During this period of uncertainty, Dixon and the men in his command would have studied weather patterns very closely, scanning the cold, late-afternoon skies for any evidence that the weather might break just long enough so they could make an attack. While Dixon waited for a calm sea, he took time to write to his old friend William Alexander to inform him about the current situation at Charleston. Alexander wrote in his memoirs, "Soon after this I received a note from Lieutenant Dixon, saying that he succeeded in getting two volunteers from the German artillery, that for two days the wind had changed to fair, and he intended to try and get out that night."[121] From a letter William Alexander penned in 1899, we know that Dixon wrote the letter on the afternoon of February 17, 1864, just hours before setting out from Breach Inlet to make history.[122] Records from the Court of Inquiry convened some nine days after the attack on Captain Pickering's vessel document his orders for the night of February 17:

> The orders to the Executive Officer and the Officer of the Deck were to keep a vigilant lookout, glasses in constant use; there were three glasses in use by the Officer of the Deck, Officer of the Forecastle and Quarter Master, and six lookouts besides; and the moment he saw anything suspicious to slip the chain, sound the gong, without waiting for orders, and send for me. To keep the engines reversed and ready for going astern, as I had on a previous occasion got my slip rope foul of the propeller by going ahead.
>
> I had the Pivot guns pivoted in broadside, the 100 pounder on the starboard side, and the eleven inch gun on the Port side; the battery all cast loose and loaded, and a round of cartridges kept in the arm chest so that two broadsides could be fired before the reception of powder from the magazine. Two shells, two canister and two grape were kept by each gun. The Quarter Gunner was stationed by the match, with the gong. Watch and lookouts armed as at Quarters. Three rockets were kept in the stands ready for the necessary signal. Two men were stationed at the slip rope, and others at the chain stopper and shackle on the spar deck. The chain was prepared for slipping by reversing the shackle, bow aft instead of forward. The pin which confined the bolt removed and a wooden pin substituted, and the shackle placed upon chain shoes for knocking the bolt out; so that all that was necessary to slip the chain was to strike the bolt with the sledge once, which broke the wooden pin, and drove the bolt across the deck, leaving the forward end of the chain clear of the shackle. I had all the necessary signals at hand, ready for an emergency. The order was to keep up 25 pounds of steam at night always, and have every thing ready for going astern instantly.[123]

A rare book now on microfilm in the Library of Congress written by a native of Charleston just twenty-four months after the attack on the *Housatonic* discusses the signal Dixon would send after the attempt had been made: "The day of the night the perilous undertaking was accomplished, the little war vessel was taken to Breach Inlet. The officer in command [Dixon] told

Lieutenant-Colonel Dantzler [commander of Battery Marshall] when they bid each other good-by, that if he came off safe he would show two blue lights."[124] At least one Federal sailor may well have seen this blue light.

At perhaps the same time that Dixon was testing his signal light and having a last-minute conversation with Colonel Dantzler at the Breach Inlet docks, the *Housatonic*'s executive officer, F. J. Higginson, was preparing to go up on deck to begin what he thought would be an uneventful evening watch. In his testimony nine days later, he provides the location of the *Housatonic* and weather conditions for the night of February 17, 1864: "The weather was clear and pleasant—moonlight, not very bright, the sea was smooth; wind about North West force 2. It was low water, and there was about 28 feet of water at her anchorage. Fort Sumter bore about West North West six miles distance. The Battery on Breach Inlet, Sullivan's Island, was the nearest land, about two and a half miles distant."[125]

At about seven o'clock, some two hours after the sun had vanished over the hazy winter horizon, Dixon took his small signal light in hand and squeezed through the forward hatch of the *Hunley*. As the dark, cold interior became illuminated from a candle Dixon had lit, seven crewmen took turns climbing down through the narrow hatches and took their places on the long wooden bench beside the crankshaft. Dixon perhaps glanced at his watch face and after a nod of his head, the gears connected to the submarine's heavy iron propeller shaft slowly started to turn, and the *Hunley* gently glided away toward the faint lights of the enemy sloop-of-war.

The events that took place aboard Dixon's submarine after leaving Breach Inlet that evening will never be known. From Alexander's writings, we can assume that soon after the *Hunley* put to sea, Dixon paused for several minutes to ballast his vessel to neutral buoyancy: "The plan was to take the bearings of the ships as they took position for the night, steer for one of them, keeping about six feet under water, coming occasionally to the surface for air and observation, and when nearing the vessel, come to the surface for final observation before striking her, which was to be done under the counter, if possible."[126]

From Alexander's eyewitness account regarding these standard operating procedures, we can conclude that once the *Hunley* had been ballasted properly, Dixon would have continued his approach on the *Housatonic*. While the submarine slowly headed out to sea toward the intended victim, Officer of the Deck John Crosby stood on the bridge of the *Housatonic* and searched the moonlit horizon for any sign of a blockade runner. At the steam sloop's loaded guns, sailors talked and joked among themselves, trying to free their minds from the boring routines associated with never-ending blockade duty. Below the vessel's frigid decks, Captain Pickering was in his cabin seated at a desk, updating a book of charts. Several bulkheads forward of Pickering's quarters stood Assistant Engineer Holihan, monitoring steam gauges in the *Housatonic*'s hot engine room, making sure that the pressure never fell below twenty-five pounds per square inch. The handwritten testimony gathered at the Court of Inquiry describes these events and reveals the vivid images that surrounded history's first successful submarine attack. We are able to read the personal observations of the *Housatonic*'s officers and crew on the night of February 17, 1864.

While Dixon monitored the dimly lit compass face in the flickering candlelight and struggled with the rudder assembly to keep the *Hunley* on course, Officer of the Deck Crosby paced the wooden deck of the *Housatonic*, searching the dark horizon for any sign of an approaching vessel. He remembers the encounter: "I took the deck at 8 P.M. on the night of February 17th. About 8:45 P.M. I saw something on the water, which at first looked to me like a porpoise, coming to the surface to blow."[127]

William Alexander writes that he considered the moonlight that evening to have been the main reason for the premature discovery of the *Hunley*: "On this night the wind had lulled, with but little sea, and although it was moonlit, Dixon, who had been waiting so long for a change of wind, took the risk of the moon light and went out. The lookout on the ship saw him when he came to the surface for his final observation before striking her."[128]

Crosby's testimony continues:

> It was about 75 to 100 yards from us on our starboard beam. The ship heading northwest by west ½ west at the time, the wind two or three points on the starboard bow. At that moment I called the Quartermaster's attention to it asking him if he saw any thing; he looked at it through his glass, and said he saw nothing but a tide ripple on the water. Looking again within an instant I saw it was coming towards the ship very fast. I gave orders to beat to quarters slip the chain and back the engine, the orders being executed immediately.[129]

As Dixon steered the submarine toward his target looming before him in the darkness, Executive Officer Higginson dashed to the *Housatonic*'s rail in an attempt to find out what all the commotion was about. He testified, "It had the appearance of a plank sharp at both ends; it was entirely on awash with the water, and there was a glimmer of light through the top of it, as though through a dead light."[130] At the same time that Higginson caught sight of Dixon's submarine boat, steadily gliding toward the hull of his vessel in the moonlight, Captain Pickering was racing on deck to find out why the ship's company had been called to quarters. He provided this testimony at the Court of Inquiry:

> I sprang from the table under the impression that a blockade runner was about. On reaching the deck I gave the order to slip, and heard for the first

time it was a torpedo, I think from the Officer of the Deck. I repeated the order to slip, and gave the order to go astern, and to open fire. I turned instantly, took my double barreled gun loaded with buck shot, from Mr. Muzzey, my aide and clerk, and jumped up on the horse block on the starboard quarter which the first Lieutenant had just left having fired a musket at the torpedo.

I hastily examined the torpedo; it was shaped like a large whale boat, about two feet, more or less, under water; its position was at right angles to the ship, bow on, and the bow within two or three feet of the ship's side, about abreast of the mizzen mast, and I supposed it was then fixing the torpedo on. I saw two projections or knobs about one third of the way from the bows. I fired at these, jumped down from the horse block, and ran to the port side of the Quarter Deck as far as the mizzen mast, singing out "Go astern Faster."[131]

Within moments after Captain Pickering had emptied both barrels of his shotgun into the *Hunley*'s forward conning tower, Lieutenant Dixon would have been shouting orders to his crew to reverse the *Hunley*'s crankshaft and back away. While terrified Union sailors leaned over the *Housatonic*'s rail and fired rifles and pistols at the strange-looking contraption, Executive Officer Higginson stood on the bridge watching the chaos by lamplight. From his sworn testimony given nine days later comes the following:

I went on deck immediately, found the Officer of the Deck on the bridge, and asked him the cause of the alarm; he pointed about the starboard beam on the water and said "there it is." I then saw something resembling a plank moving towards the ship at a rate of 3 or 4 knots; it came close along side, a little forward of the mizzen mast on the starboard side. It then stopped, and appeared to move off slowly. I then went down from the bridge and took the rifle from the lookout on the horse block on the starboard quarter, and fired it at this object.[132]

As desperate sailors on deck tried to stop the approach of the infernal machine with small-arms fire, Assistant Engineer Mayer was in the engine room, trying to engage the gears to the *Housatonic*'s huge propeller and move the stationary warship astern. From his testimony, "Three bells were struck a few seconds after I got there, the engine was immediately backed, and had made three or four revolutions when I heard the explosion, accompanied by a sound of rushing water and crashing timbers and metal. Immediately the engine went with great velocity as if the propeller had broken off. I then throttled her down, but with little effect. I then jumped up the hatch, saw the ship was sinking and gave orders for all hands to go on deck."[133]

While Assistant Engineer Mayer struggled in the hot, dimly lit engine room in an attempt to move the *Housatonic* astern, Ensign Charles Craven spring from his bunk and rushed on deck. He testified:

I heard the Officer of the Deck give the order "Call all hands to Quarters." I went on deck and saw something in the water on the starboard side of the ship, about 30 feet off, and the Captain and the Executive Officer were firing at it. I fired two shots at her with my revolver as she was standing towards the ship as soon as I saw her, and a third shot when she was almost under the counter, having to lean over the port to fire it.

    I then went to my division, which is the second, and consists of four broadside 32 pounder guns in the waist, and tried with the Captain of number six gun to train it on this object, as she was backing from the ship, and about 40 or 50 feet off then. I had nearly succeeded, and was almost about to pull the lock string when the explosion took place. I was jarred and thrown back on the topsail sheet bolts, which caused me to pull the lock string, and the hammer fell on the primer but without sufficient force to explode it. I replaced the primer and was trying to catch sight of the object in order to train the gun again upon it, when I found the water was ankle deep on deck by the main mast. I then went and assisted in clearing away the second launch.[134]

From numerous individuals who gave testimony at the Court of Inquiry, we can conclude that when the explosive charge was detonated, Dixon was much closer to the *Housatonic*'s hull than he would have wanted. As the little submarine undoubtedly shuddered and pitched in the reverberating shock waves triggered by the powerful blast, Federal sailors, trapped below deck, desperately struggled in freezing seawater to reach an open hatch. With water then gushing through the gaping hole blown out by the torpedo, the hull of the doomed *Housatonic* slowly rolled to port as freezing water rose beneath the decks. In an attempt to escape the frigid winter waters then rising around them, officers and men scrambled into lifeboats and onto the ship's rigging that remained above the surface.

    The testimony of Acting Master Joseph Congdon describes the situation aboard the *Housatonic* immediately following the explosion. "I drew my revolver, but before I could fire, the explosion took place. I immediately went forward and ordered the launches to be cleared away, supposing the captain and Executive Officer had both been killed by the explosion. The ship was sinking so rapidly, it seemed impossible to get the launches cleared away, so I drove the men up the rigging to save themselves. After I got into the rigging I saw two of the boats had been cleared away, and were picking up men who were overboard. As soon as I saw all were picked up, I sent one of the boats to the 'Canandaigua' for assistance."[135]

    From the logbook of the USS *Canandaigua* then on station near the *Housatonic* on the night of February 17, Capt. Joseph Green penned the following entry: "February 17th, 1864. At 9:20 P.M. discovered a boat pulling towards us. Hailed her and found her to be from the 'Housatonic.' She reported the

'Housatonic' sunk by a torpedo. Immediately slipped our chain and started for the scene of danger, with the 'Housatonic's' boat in tow. At same time sent up three rockets and burned Coston signals number 82. At 9:30 P.M. picked up another boat from the 'Housatonic,' with Captain Pickering on board. At 9:35 arrived at the 'Housatonic' and found her sunk. Lowered all boats, sent them alongside, and rescued the officers and crew, clinging to the rigging."[136]

From sworn testimony, it appears that Captain Singer's 135-pound torpedo had detonated well below the *Housatonic*'s water line, accounting for the agreed fact that no one heard a terrific explosion. Officer of the Deck John Crosby testified: "The explosion started me off my feet, as if the ship had struck hard on the bottom. I saw fragments of the wreck going up into the air. I saw no column of water thrown up, no smoke and no flame. There was no sharp report, but it sounded like a collision with another ship."[137] Since the blast took place several feet beneath the surface (as was originally intended, judging from both Chief Engineer Tomb's and William Alexander's postwar writings),[138] the force of the concussion would have been entirely directed against the *Housatonic*'s wood hull, which was the path of least resistance.

From testimony given by numerous survivors, the *Hunley* was quite close to the *Housatonic* when the torpedo exploded. With the submarine so close, the reel holding the line that was to pull the torpedo's detonating pin probably jammed or perhaps malfunctioned in some other way.

A fascinating observation that has remained a subject of speculation and debate for more than a century came to light during the Court of Inquiry. Seaman Robert Fleming, the lookout posted at the starboard cathead, remembered what he saw after he had climbed the *Housatonic*'s rigging that remained above the water: "When the 'Canandaigua' got astern, and was lying athwart, of the 'Housatonic,' about four ship lengths off, while I was in the fore rigging, I saw a blue light on the water just ahead of the 'Canandaigua,' and on the starboard quarter of the 'Housatonic.'"[139]

From a report filed just forty-eight hours after the attack, Fleming may well have seen the faint beams from Dixon's blue signal light. The commander of Battery Marshall, Lieutenant Colonel Dantzler, reinforces this observation, for in his official report on the incident, he clearly states that the agreed signal from Lieutenant Dixon was both "observed and answered."[140] Whether or not seaman Fleming actually saw Dixon's signal light will perhaps never be known, for after successfully sinking the federal sloop-of-war *Housatonic*, Lt. George E. Dixon, his submarine, and the entire crew vanished, and were never heard from again.

1. Texan Edgar Collins Singer was in command of the Singer Secret Service Corps (also known from surviving documentation as the "Singer Submarine Corps" and "Singer's Torpedo Company").
From Hill, "Texans Gave World First Successful Submarine Torpedo."

2. Port Lavaca, Texas, mid-nineteenth century.
Courtesy of the Calhoun County Museum.

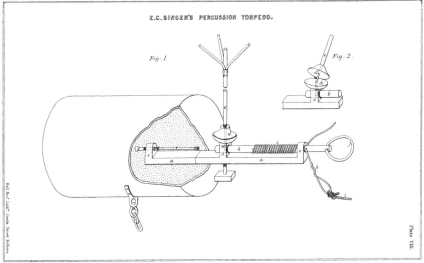

3. Sketch of the Port Lavaca waterfront just prior to the war in 1858, *top*.
Courtesy of the Calhoun County Museum.

4. Diagram of Edgar Singer's first underwater "percussion torpedo."
More Union warships were damaged or sunk by Singer torpedoes during the war
than any other variety manufactured in the Confederacy.
From Von Scheliha, *A Treatise on Coast-Defence*, 228.

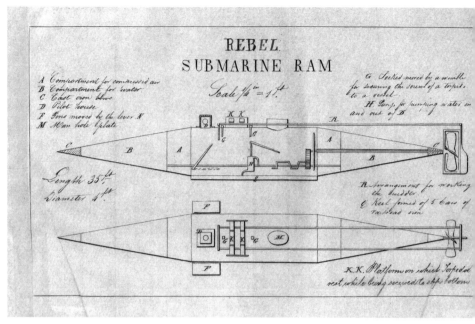

5. Diagram of Confederate privateer submarine *Pioneer* drawn by Fleet Engineer William Shock soon after the fall of New Orleans. The scuttled vessel was discovered soon after the collapse of the city and dragged to shore by Union sailors.
Source: Diagram attached to letter from J. H. Shock to Asst. Naval Sec. G. W. Fox dated January 1864, in Letters Received by the Secretary of the Navy from Officers below the Rank of Commander, Record Group 45, National Archives.

6. James McClintock, *below*, and Horace Hunley, *right*.
Source: Photograph Archives, Naval Historical Center.

7. Gen. Dabney H. Maury.
Source: Library of Congress.

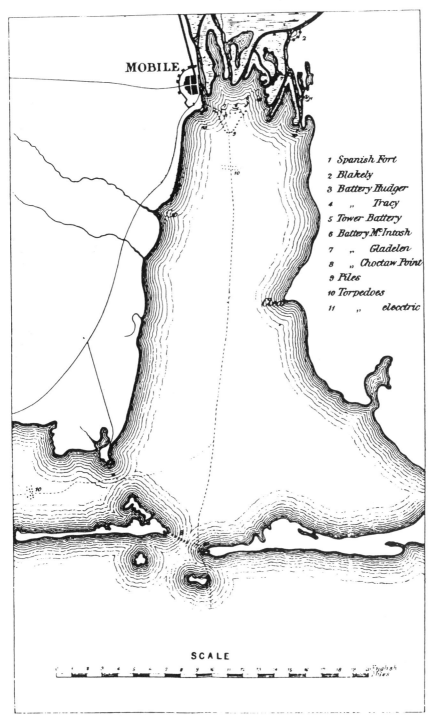

8. Mobile Bay. Adapted from "A Treatise on Coast Defence—1868,"
Source: Library of Congress.

9. USS *Baron DeKalb*, first Union warship sunk by a Singer mine, in the Yazoo River, Mississippi, on July 13, 1863, *top*.
Source: Naval Historical Center.

10. Park and Lyons machine shop, at the corner of Water and State Streets in Mobile, where both the *Hunley* submarine and Singer torpedoes were fabricated during the war. It was still serving as a machine shop when this photograph was taken in 1960, but the building has since been destroyed.
Source: Naval Historical Center.

The Confederate States, For _____ To _Whitney & Brennan_ **Dr.**

DATE.	ORDNANCE SERVICE.			REMARKS.
June 30th	For 20. E C Singers Torpedoes. made under Contract with Genl Buckner	$150 00	$3000 00	

Received, Mobile, Ala., June 30th, 1863, from Capt Henry Myers the sum of Three Thousand dollars and _____ cents, in full payment of the above account.

$3000   (Signed Duplicates.)   Whitney & Brennan

11. Receipt for payment of three thousand dollars for twenty Singer torpedoes.
Source: National Archives.

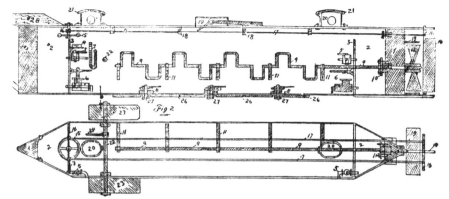

LONGITUDINAL ELEVATION IN SECTION AND PLAN VIEW OF THE CONFEDERATE SUBMARINE BOAT HUNLEY.
From Sketches by W. A. Alexander.

No. 1. The Bow and Stern Castings. No. 2. Water ballast tanks. No. 3. Tank bulkheads. No. 4. Compass. No. 5. Sea cocks. No. 6. Pumps. No. 7. Mercury gauge. No. 8. Keel ballast stuffing boxes. No. 9. Propeller shaft and cranks. No. 10. Stern bearing and gland. No. 11. Shaft braces. No. 12. Propeller. No. 13. Wrought ring around propeller. No. 14. Rudder. No. 15. Steering wheel. No. 16. Steering lever. No. 17. Steering rods. No. 18. Rod braces. No. 19. Air box. No. 20. Hatchways. No. 21. Hatch covers. No. 22. Shaft of side fins. No. 23. Cast iron keel ballast. No. 27. Bolts. No. 28. Butt end of torpedo boom. No. 23. Side fins. No. 24. Shaft lever. No. 25. One of the crew turning propeller shaft. No. 31. Keel ballast

CSS H. L. Hunley (1863-64). This duel view was drawn by William A. Alexander, who directed her construction. *U.S. Naval Historical Center*

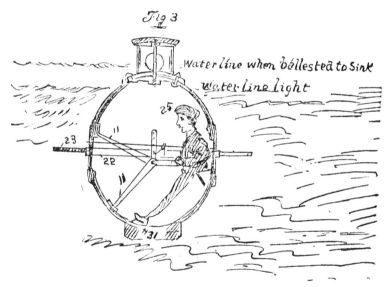

AMIDSHIPS SECTIONS OF THE HUNLEY SUBMARINE.
From Sketches by W. A. Alexander.

12. Diagram of the *Hunley* by William Alexander, *top*.
Source: Naval Historical Center.

13. Sketch of the *Hunley*'s interior by William Alexander, 1902.
Source: Naval Historical Center.

14. Confederate receipt for eighty Singer torpedoes issued at about the same time that the *Hunley* was being tested in the Mobile River.
Source: National Archives.

15. Charleston's commanding general P. G. T. Beauregard met with Singer representatives in early August and immediately agreed that the *Hunley* should be sent to Charleston to attack the Federal ironclads blockading the harbor.
Source: National Archives.

16. A wartime view of Charleston's East Bay Street, *above*.
Source: National Archives.

17. Nineteenth-century map of Charleston Harbor.
Source: Stock art.

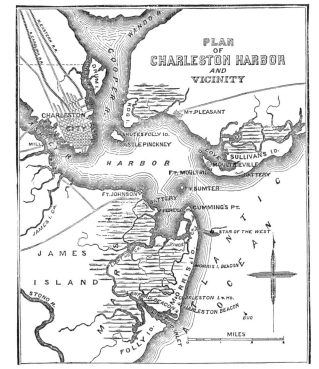

18. The Federal ironclad steamer USS *Ironsides* was the *Hunley*'s main target during its first few weeks in Charleston, *top*.
Source: Naval Historical Center.

19. Warm weather often brought crews of the ironclads on deck to escape the sweltering heat below.
Source: Library of Congress.

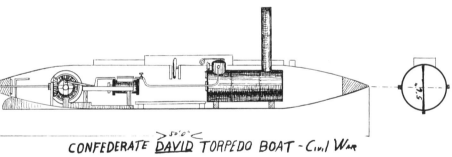

20. Federal gun battery on Morris Island, *top*.
Source: Library of Congress.

21. Diagram of interior of a *David*-class torpedo boat with boiler, steam engine, flywheel, and propeller shaft clearly indicated. With internal ballast tanks, this craft could be submerged up to the base of the smokestack; thus, several modern historians refer to this class of torpedo boat as a submarine.
Source: Naval Historical Center.

22. Rear Admiral John A. Dahlgren, commander of the South Atlantic Blockading Squadron. Source: Library of Congress.

23. William Alexander (1902), who served as second in command under Lt. George Dixon.
Source: Mobile City Museum, Mobile, Alabama.

(No. 40.)

SPECIAL REQUISITION.

For *Nine Gray Jackets, three to be trimmed with Gold braid*

I certify that the above Requisition is correct; and that the articles specified are absolutely requisite for the public service, rendered so by the following circumstances: *that the men for whom they are ordered, are on Special Scout Service & that it is necessary that they be clothed in the Confederate Army uniform* H. L. Hunley

Capt G. I. Crafts A.Q. Quartermaster, C. S. Army, will issue the articles specified in the above requisition. *By Cmd Brig Genl Ripley*
Commanding

Received at *Charleston* the *21st* of *August* 186*3* of *Capt G. I. Crafts A.Q.* Quartermaster, C. S. Army, *The Jackets specified above*

in full of the above requisition.
(SIGNED DUPLICATES.) H. L. Hunley
Capt

---

24. As a high-ranking member of the Singer Secret Service Corps, Horace Hunley received the rank of captain. "H. L. Hunley Capt." filed this special requisition for "Nine Gray Jackets" for the crew of the submarine on August 21, 1863.
Source: Unfiled Papers and Slips Belonging in Confederate Compiled Service Records.

25. A huge Federal siege cannon nicknamed "the swamp angel" fired the first shots into Charleston at 1:30 a.m. on August 22, 1863.
Source: Frank Leslie's *Illustrated News*, National Archives.

SINKING TORPEDOES IN CHARLESTON HARBOR BY MOONLIGHT.—From a Drawing by an English Artist.—[See Page 375.]

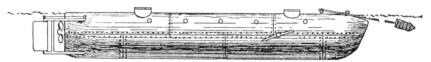

THE CONFEDERATE SUBMARINE BOAT WHICH SANK THE UNITED STATES STEAMSHIP "HOUSATONIC" IN CHARLESTON HARBOR DURING THE CIVIL WAR.

Three different crews were drowned in this boat before she accomplished her purpose. As shown in the two upper diagrams, her propeller and two forward paddles were worked with a shaft propelled by eight men. The steersman, in front, discharged the torpedo, shown in the lower view.

26. Sinking torpedoes into Charleston harbor by moonlight. Large electrically detonated torpedoes hidden just below the surface kept Federal warships and ironclads at bay, *top*.
Source: *Harper's Weekly*, June 13, 1863.

27. *Hunley* diagram by Lt. Charles Hasker (1897), depicting how he remembered the configuration of the spar torpedo. Prior to the excavation of the submarine in 2000, it was thought that this was the way that the torpedo was delivered, but the spar assembly was actually attached to the lower bow.
Source: *McClure's Magazine*, January 1899, Library of Congress.

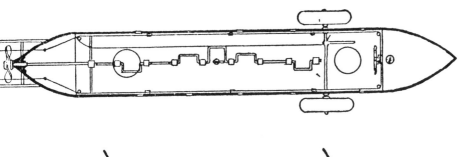

28. A stripped and abandoned *David*-class torpedo boat after the war, *top*.
Source: Library of Congress.

29. Diagrams of the *Hunley* by former Confederate naval officer Charles Hasker, summer 1897. Although incorrect on some points, these diagrams are the only known nineteenth-century depictions of the *Hunley* to show the topside viewports.
Source: *McClure's Magazine*, January 1899, Library of Congress.

30. Wartime sketch by Conrad Chapman of Battery Marshall from the far side of Breach Inlet (1863), *top*.
Source: Military History Institute, Carlisle, Pennsylvania.

31. Sketch of a nighttime bombardment of Fort Sumter showing the fort illuminated by a calcium light shining from an enemy gun emplacement on Morris Island. This scene would have been a common sight to Lieutenant Dixon and the crew of the *Hunley* as they set out from Mount Pleasant to attempt a nocturnal attack against the ironclads at anchor outside the harbor.
Source: Military History Center, Carlisle, Pennsylvania.

THE MERCURY.

BY R. B. RHETT, JR.

OFFICE NO. 484 KING-STREET, CHARLESTON.

THE DAILY MERCURY, ten cents per copy, $20 per annum.
THE TRI-WEEKLY MERCURY, issued on Tuesdays, Thursdays and Saturdays, ten cents per copy, $10 per annum.
ADVERTISEMENTS, Two Dollars per square of 13 lines.

MONDAY, NOVEMBER 9, 1863.

LAST HONORS TO A DEVOTED PATRIOT.—The remains of Captain HORACE L. HUNLEY were yesterday interred in Magnolia Cemetery. His body was followed to the grave by a military escort, and a large number of citizens.

The deceased was a native of Tennessee, but for many years past has been a resident of New Orleans.

Possessed of an ample fortune, in the prime of manhood—for he was only thirty-six at the time of his death—with everything before him to make life attractive, he came to Charleston, and voluntarily joined in a patriotic enterprise which promised success, but which was attended with great peril. Though feeling, as appears from the last letter which he wrote to his friends, a presentiment that he would perish in the adventure, he gave his whole heart, undeterred by the foreboding, to the undertaking, declaring that he would gladly sacrifice his life in the cause. That presentiment has been mournfully fulfilled. Yet who shall call that fate a sad one, which associates the name of its victim with those of his country's most unselfish martyrs?

32. Article written after burial of Captain Hunley in Magnolia cemetery. Source: Charleston *Mercury*, November 9, 1863, Library of Congress.

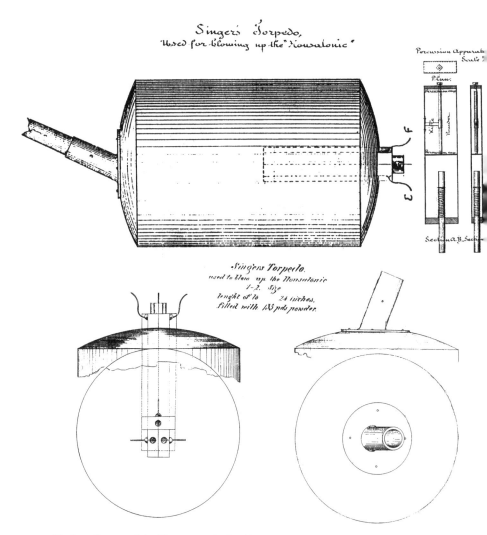

33. Wartime diagram of the Singer torpedo that blew up the *Housatonic* (1864). Source: National Archives.

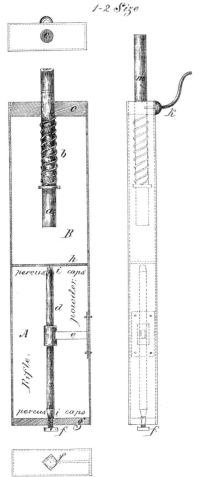

34. This captured diagram of the Singer torpedo fuse shows the spring-loaded detonator on the *Hunley*'s torpedo.
Source: National Archives.

35. The 1,240-ton sloop-of-war *Housatonic* was the first warship in history to be sunk by an enemy submarine. This act would not be repeated until World War One more than fifty years later.
Source: Naval Historical Center.

36. Description of the USS *Housatonic* listing armaments and boilers.
Source: Naval History Center.

...kee advance was made with the intention of a general attack.

LIEUT. DIXON.—The Mobile *Register* learns that Lieut. Dixon, who blew up the Housetonic, on the night of the 13th February, was either lost or captured by the enemy on his return. A dispatch was received in Mobile on the morning of March 1st, stating that he was last seen heading in the direction of Fort Sumter, between which and himself were two lines of the enemy's picket boats. It was expected that his fate would be ascertained by the next flag of truce.

37. Judging from this short entry in the March 15 edition of the Fayetteville (North Carolina) *Weekly Intelligencer*, the name of Lt. George Dixon echoed across the Confederacy during the weeks that followed the sinking of the *Housatonic*, top.
Source: Library of Congress.

38. In July 2000, the huge crane barge *Karlissa-B* was towed from the Dominican Republic and moored next to the *Hunley*. During the weeks that followed underwater archaeologists and professional divers worked around the clock to free the *Hunley* from its century-old grave at the bottom of the sea.
Photograph by the author.

39. Project divers and archaeologists were lowered to the work site in this metal enclosure, which was positioned some twenty feet above the ocean surface. Here the author awaits the crane operator's signal that he is ready to commence the descent.
Photo by unknown crew member.

40. With the *Hunley* securely cradled beneath the metal truss, crane operator Jenkins Montgomery guides the submarine toward the waiting barge.
Photo by the author.

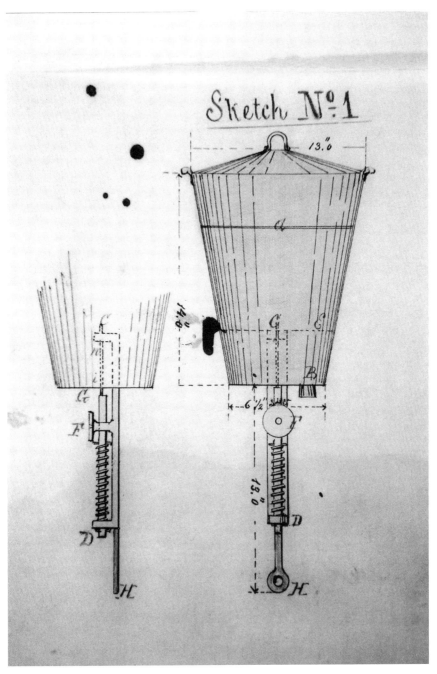

41. In early 1864 the Singer group submitted patent diagrams for an improved underwater contact mine. This diagram is actually a Federal blueprint based on this new configuration of a Singer mine that was recovered near the mouth of Mobile Bay during the summer of 1864. Source: National Archives.

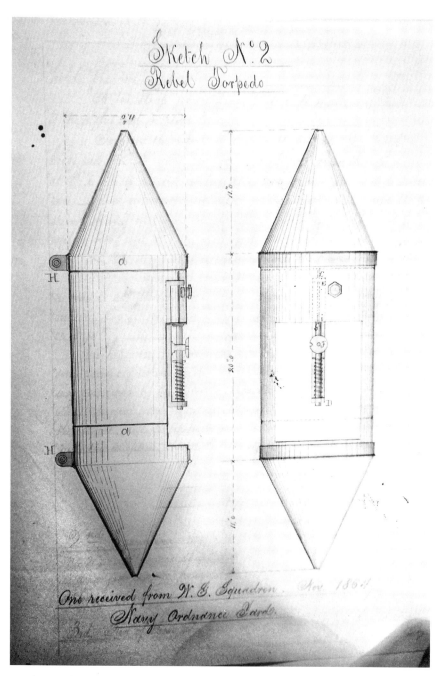

42. Diagram of a Singer torpedo recovered by the Federals in Mobile Bay. With its pointed ends, this mine would have been best suited for deployment in areas with strong currents. Source: National Archives.

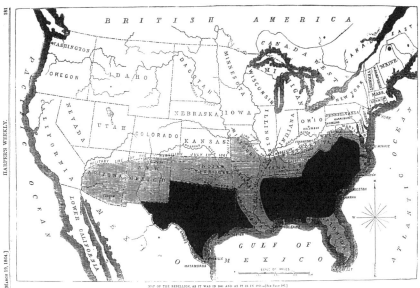

43. The Spotswood Hotel in Richmond served as the headquarters for the Singer Secret Service Corps during the last two years of the war, *top*.
Source: Library of Congress.

44. Map showing how the Confederacy had shrunk since 1861.
Source: *Harper's Weekly*, 1864, Library of Congress.

45. This segment of an early 1864 Confederate muster roll, naming all those engaged in operations in Mobile Bay, lists James McClintock, Greenleaf Andrews, Baxter Watson, and John A. King as being members of Singer's Secret Service Corps under the authority of the War Department.
Source: National Archives.

46. The Federal ironclad monitor *Tecumseh* was sunk by a Singer mine with great loss of life on August 5, 1864.
Source: *Harper's Weekly*, Library of Congress.

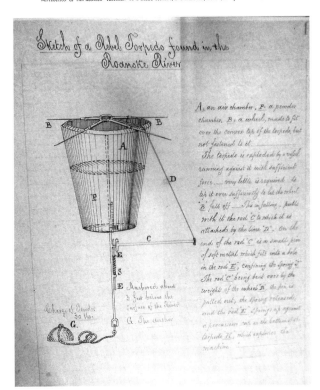

47. Diagram produced after examining a recovered Singer mine that John Fretwell had anchored in the Roanoke River, North Carolina, in late 1864.
Source: National Archives.

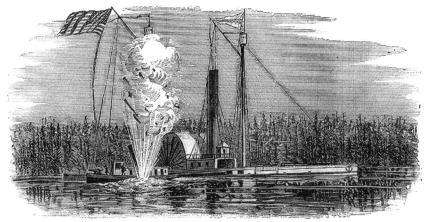

WRECK OF THE "OTSEGO," AND THE EXPLOSION OF THE TUG "BAZLEY" IN THE ROANOKE RIVER, DECEMBER 10, 1864.

48. Engraving depicting the sinking of the Federal tug *Bazley* in the Roanoke River, North Carolina. The Singer mine pictured next to the engraving (with accompanying description of how it functioned) was discovered near the wreck soon after the disaster. Members of the Singer Secret Service Corps remained active in the Confederate war effort until virtually the last guns fell silent. Source: *Harper's Weekly*, January 21, 1865, Library of Congress.

49. *Above and right.* Of the five Singer Secret Service Corps members who purchased shares in the *Hunley*, only Dunn, Braman, and Singer survived the war (Whitney died of pneumonia shortly after military authorities seized the *Hunley* in late August, and Hunley drowned in the submarine on October 15, 1863). These parole papers signed by the surviving members prove that they were all on duty in Texas when the war ended in 1865.
Source: National Archives.

No. 2

I, the undersigned, Prisoner of War, belonging to the Army of the Trans-Mississippi Dept. having been surrendered by Genl. E. K. Smith, C. S. A., Commanding said Department, to Maj. Gen. E. R. S. Canby, U. S. A., Commanding Army and Division of West Mississippi, do hereby give my solemn **Parole of Honor**, that I will not hereafter serve in the Armies of the Confederate States, or in any military capacity whatever, against the United States of America, or render aid to the enemies of the latter, until duly exchanged, or otherwise released from the obligations of this Parole by the authorities of the Government of the United States.

Done at Lavasa, Texas this 8th day of July, 1865.

E. C. Singer
Comdg Singer Special Service Company

M. Bailey
Maj. 7th U.S.C.T. & Prov. Mar.

The above named officer will not be disturbed by the United States authorities, as long as he observes his Parole, and the laws in force where he resides.

M. Bailey
Maj. 7th U.S.C.T.
Prov. Mar. Lavaca, Texas

50. The Port Lavaca Masonic Lodge as it appeared in 2004. All of the founding members of the Singer Secret Service Corps were brothers in this lodge prior to the outbreak of the war. Photo by the author.

51. From as early as 1980 Clive Cussler led several expeditions to locate the final resting place of the Confederate submarine *H. L. Hunley*. He and his team succeeded in 1995.
Photo by Tom Story.

Losing the *Hunley*   4

As the first faint rays of sunlight slowly appeared over the Atlantic's cold horizon on the morning of February 18, shivering pickets tossed the last few pieces of driftwood on dying embers and continued to scan the gray waves offshore for any sign of Lieutenant Dixon and the submarine boat under his command.[1] While Lieutenant Colonel Dantzler and the soldiers stationed at Battery Marshall presumably wondered about the whereabouts of the overdue *Hunley*, Capt. Joseph Green of the USS *Canandaigua* was hastily putting to paper the following report:

> U.S.S. "Canandaigua," off Charleston, S.C., February 18, 1864. Commander Rowan. Sir: I have respectfully to report that a boat belonging to the 'Housatonic' reached this ship last night at about 9:20, giving me information that vessel had been sunk at 8:45 p.m. by a rebel torpedo craft. I immediately slipped our cable and started for her anchorage, and on arriving near it, at 9:35, discovered her sunk with her hammock netting under water; dispatched all boats and rescued from the wreck 21 officers and 129 men.
>
> Captain Pickering was very much, but not dangerously, bruised and one man is slightly bruised. I have transferred to the "Wabash," 8 of her officers and 49 men, on the account of the limited accommodations on board of this vessel. Very respectfully, your obedient servant, J. F. Green, Captain.[2]

Captain Pickering, who had been badly injured, delegated to Executive Officer F. J. Higginson the responsibility of reporting the disaster to Admiral Dahlgren, who was then at Port Royal, South Carolina:

> U.S.S. "Canandaigua," off Charleston, S.C., February 18, 1864. Sir: I have the honor to make the following report of the sinking of the "U.S.S. Housatonic," by a rebel torpedo off Charleston, S.C. on the evening of the 17th instant. About 8:45 p.m. the officer of the deck, Acting Master J. K. Crosby, discovered something in the water about 100 yards from and moving toward the ship. It had the appearance of a plank moving on the water. It came directly toward the ship, the time from when it was first seen till it was close alongside being about two minutes.
>
> During this time the chain was slipped, engine backed, and all hands called to quarters. The torpedo struck the ship forward of the mizzenmast, on the

starboard side, in a line with the magazine. Having the after pivot gun pivoted to port we were unable to bring a gun to bear upon her. About one minute after she was close alongside the explosion took place, the ship sinking to stern first and heeling to port as she sank. Most of the crew saved themselves by going into the rigging, while a boat was dispatched to the "Canandaigua." This vessel came gallantly to our assistance and succeeded in rescuing all but the following named officers and men, viz, Ensign E. C. Hazeltine, Captain's Clerk C. O. Muzzey, Quartermaster John Williams, Landsman Theodore Parker, Second-Class Fireman John Walsh. The above officers and men are missing and are supposed to have been drowned. Captain Pickering was seriously bruised by the explosion and is at present unable to make a report of the disaster.[3]

Battery Marshall's commanding officer Lt. Col. O. M. Dantzler reported on the confused events that led to the Confederate discovery of the *Housatonic*'s destruction:

Head Quarters, Battery Marshall, Sullivan's Island. Feb. 19th, 1864. Lieutenant: I have the honor to report that the torpedo boat stationed at this post went out on the night of the 17th instant (Wednesday) and has not yet returned. The signals agreed upon to be given in case the boat wished a light to be exposed at this post as a guide for its return were observed and answered. An earlier report would have been made of this matter, but the Officer of the Day for yesterday was under the impression that the boat had returned, and so informed me. As soon as I became apprised of the fact, I sent a telegram to Captain Nance assistant adjutant-general, notifying him of it. Very respectfully Col. Dantzler.[4]

According to Lieutenant Colonel Dantzler's official report of the incident, the *Hunley* may indeed have survived its attack on the *Housatonic* and remained seaworthy just long enough for Lieutenant Dixon to flash a message to light the signal beacon at the mouth of Breach Inlet. Since the colonel's report mentioned nothing regarding the sinking of an enemy vessel, the weather conditions for February 18 and 19 may have been too hazy for anyone atop Battery Marshall's wooden watchtower to observe the *Housatonic*'s exposed masts. While news regarding the overdue *Hunley* circulated at Confederate military headquarters in Charleston, Admiral Dahlgren issued the following orders from his flagship then anchored at Port Royal:

Flag-Steamer "Philadelphia," Port Royal Harbor, S.C., February 19th, 1864. The "Housatonic" has just been torpedoed by a rebel "David," and sunk almost instantly. It was at night and water smooth. The success of this undertaking will, no doubt, lead to similar attempts along the whole line of the blockade. If vessels on the blockade are at anchor they are not safe, particularly in smooth water, without outriggers and hawsers stretched around with rope netting dropped into the water. Vessels on inside blockade had better take post outside at night and keep underway, until these preparations are completed.

All the boats must be on the patrol when the vessel is not in movement. The commanders of vessels are required to use their utmost vigilance—nothing less will serve. I intend to recommend to the Navy Department the assignment of a large reward as prize money to crews of boats or vessels who shall capture, or beyond doubt destroy, one of these torpedoes. Admiral Dahlgren.[5]

On the morning after these orders were issued, Captain Green and a small contingent of officers from the USS *Canandaigua* set course for the wreck site of the *Housatonic* and examined the damage firsthand. Green reported what he discovered while Confederate sentries closely observed him from shore : "U.S.S. 'Canandaigua,' February 20, 1864. Sir: I have examined the wreck of the 'Housatonic' this morning and find her spar deck about 15 feet below the surface of the water. The after part of her spar deck appears to have been entirely blown off. Her guns, etc., on the spar deck, and probably a good many articles below deck, can, in my opinion, be recovered by the employment for the purpose of the derrick boat and divers."[6]

While Captain Green and several Federal naval officers examined the *Housatonic* wreck, the assumption that Lieutenant Dixon had successfully sunk an enemy ship on the night of February 17 appears to have taken hold at Battery Marshall. Several tugs and barges were seen at work around the exposed masts of a sunken blockader, so that the wreck must have been a victim of the missing *Hunley*. Within hours after the sighting, Lieutenant Colonel Dantzler sent word of the discovery to General Ripley at Mount Pleasant, and by early afternoon, General Jordan at Charleston headquarters received the following report: "Mount Pleasant, February 20th, 1864, Brig. General Jordan, Chief of Staff: Lt. Col. Dantzler reports a gunboat sunk off Battery Marshall, smoke stack and rigging visible, a tug boat and barge are around her, supposed to be the 'Flambeau.' Another has not been seen since Wednesday night and it may be that she was blown up by the missing torpedo boat."[7]

As word of the sinking of a Federal blockader spread through the Charleston garrison, General Beauregard wasted little time notifying his superiors in Richmond: "To General S. Cooper, Inspector-General, Confederate States Army, Richmond, Va. Charleston, S.C., February 21, 1864, General: A gunboat sunk off Battery Marshall. Supposed to have been done by Mobile torpedo boat under Lieutenant George E. Dixon, Company A, Twenty-first Alabama Volunteers, which went out for that purpose, and which I regret to say has not been heard of since. G. T. Beauregard."[8]

On the night of February 26, some six days after sentries at Battery Marshall first observed the *Housatonic*'s masts, a Federal picket vessel strayed too close to Fort Sumter and was captured by a Confederate patrol boat attached to the CSS *Indian Chief*. While interrogating the six Federals found aboard, they revealed that the *Housatonic* was the warship sunk some nine nights earlier and that five crewmen had been killed in the attack.⁹ General Beauregard then quickly sent the following telegram to General Cooper at the Confederate capital: "Charleston, S.C., February 27, 1864. General S. Cooper: Prisoners report that it was the U.S. ship of war Housatonic, 12 guns, which was sunk on night 17th instant by the submarine torpedo boat, Lieutenant Dixon, of Alabama, commanding. There is little hope of safety of that brave man and his associates, however as they were not captured. G. T. Beauregard."¹⁰

The enemy's long-range siege guns constantly shelled the city, so news of the sinking of an enemy ship was reason for rejoicing, and military headquarters appears to have purposely suppressed the fact that the *Hunley* did not return after the attack. As long as Admiral Dahlgren was unaware that the torpedo boat that sunk his blockader was lost in the attack, he would be forced to waste time protecting his fleet against a second attack that would never come.

On the last day of February 1864, the *Charleston Mercury* printed a glowing account of the *Housatonic*'s destruction that was reprinted in newspapers throughout the Confederacy in the days that followed:

> Monday, February 29, 1864. Siege Matters—Two Hundred and Thirty Fifth Day. The news this morning from our immediate vicinity is quite as cheering as that which is echoed along the wires from the far off battlefields of Georgia and the Southwest. An official dispatch was received from Colonel Elliott at Fort Sumter, on Saturday, conveying the gratifying news that one of our picket boats, commanded by boatswain Smith, had captured a Yankee picket boat containing one officer and five men. The prisoners have arrived in the city. Their accounts of the success of the pioneer of our fleet of torpedo boats are really exhilarating. They state that the vessel sunk off the harbor on the night of the 16th and reported lost in a gale, was the U.S. steamer Housatonic, carrying 12 guns and three hundred men, and that she was blown up by our torpedo boat.
>
> This fine and powerful vessel was sunk in three minutes. The whole stern of the steamer was blown off by the explosion. All of the crew of the Housatonic are said to have been saved, except five—two officers and three men—who are missing and are supposed to be drowned. As a practical and important result of the splendid achievement, the prisoners state that the wooden vessels of the blockading squadron now go far out to sea every night, being afraid of the risk of riding at anchor in any portion of the outer harbor.

The torpedo boat that accomplished this glorious exploit was under the command of Lieutenant Dixon. We are glad to be able to assure out readers that the boat and crew are now safe. Since our last report the shelling of the city has been maintained by the enemy with undiminished vigor. Two hundred and six shells have been thrown—one hundred and six on Saturday and one hundred on Sunday. At midnight the bombardment was still going on very briskly, one shell being thrown every five minutes. The report form Fort Sumter is "All Quiet."

According to a letter written by Harriet Middleton of Charleston to her sister, we can conclude that the nocturnal activities of Lieutenant Dixon and the crew of the *Hunley* had, in the days following news of the attack, reached mythic proportions. These fanciful excerpts regard two additional *Hunley* victories, as well as news that the base of operations for the submarine had been moved some forty miles farther up the coast to Georgetown: "We can't get at the truth of the Fish boat story. Some say she has never returned—while the generals believe she came ashore safe somewhere near Georgetown. She is said to have sunk two vessels besides the 'Housatonic'—one a transport loaded with troops. This time she did not dive, but attacked the enemy on the surface of the water—They seeing her all the time, and firing grape and canister at her without any effect. Torpedoes have been sent on to her at Georgetown to start on another expedition."[11] Apparently, fanciful rumors regarding further nocturnal escapades and additional victories of Lieutenant Dixon and his crew were flying around Charleston at a dizzying speed. We can assume that military authorities were purposely circulating false statements regarding the *Hunley*'s survival and escape to Georgetown for both morale and security reasons.[12]

William Alexander wrote later of his reaction to the story we assume he read in the *Mobile Advertiser* some days later: "Next came the news that on February 17th the submarine torpedo boat 'Hunley' had sunk the United States sloop-of-war 'Housatonic' outside the bar off Charleston, S.C. As I read, I cried out with disappointment that I was not there. Soon I noted that there was no mention of the whereabouts of the torpedo boat. I wired General Jordan daily for several days, but each time came the answer, 'No news of the torpedo boat.' After much thought, I concluded that Dixon had been unable to work his way back against wind and tide, and been carried out to sea."[13]

In the days and weeks that followed, it appears that most every Confederate newspaper in the South (judging from those currently on file at the Library of Congress), copied the February 29 Charleston news article. From the office of Jefferson Davis to the backwater garrisons scattered along the Texas coast, news that a phantom Confederate torpedo boat attacked and sank an enemy blockader outside Charleston Harbor on the night of February 17 gave a much-needed shot in the arm to southern morale. From the following article in the March 15, 1864, edition of the *Fayetteville* (North Carolina) *Weekly Intelligencer* we can conclude that citizens throughout the blockaded South anxiously awaited news of the fate of Dixon and his daring crew: "The Mobile Register learns that Lieutenant Dixon, who blew up the Housatonic on the night of the 18th February was either lost or captured by the enemy on his return. A dispatch was received in Mobile on the morning of March 1st, stating that he was last seen heading in the direction of Fort Sumter, between which and himself were two lines of the enemy's picket boats. It was expected that his fate would be ascertained by the next flag of truce."

Judging from the Federal newspapers currently on file at the Library of Congress, northern citizens also considered the event extremely newsworthy, for virtually every paper examined from the period in question reported the attack in great detail. The northern press in most instances vilified the encounter as an unchivalrous rebel act of desperation.

While distorted rumors regarding additional Hunley victories and escape northward to Georgetown spread throughout the garrison troops and citizens of Charleston, Captain Pickering and the officers and men under his command were questioned at a Court of Inquiry, which reached the following conclusions:

> First. That the U.S.S. "Housatonic" was blown up and sunk by a rebel torpedo craft on the night of February 17 last, about 9 o'clock p.m., while lying at an anchor in 27 feet of water off Charleston S.C. . . .
>
> Second. That between 8:45 and 9 o'clock p.m. on said night an object in the water was discovered almost simultaneously by the officer of the deck and the lookout stationed at the starboard cathead, on the starboard bow of the ship, about 75 or 100 yards distant, having the appearance of a log. . . .
>
> Third. That the strange object approached the ship with a rapidity precluding a gun of the battery being brought to bear upon it, and finally came in contact with the ship on her starboard quarter.
>
> Fourth. That about one and a half minutes after the first discovery of the strange object the crew were called to quarters, the cable slipped, and the engine backed.
>
> Fifth. That an explosion occurred about three minutes after the first discovery of the object, which blew up the after part of the ship, causing her to sink immediately after to the bottom, with her spar deck submerged.
>
> Sixth. That several shots from small arms were fired at the object while it was alongside or near the ship before the explosion occurred.
>
> Seventh. That the watch on deck, ship, and ship's battery were in all respects prepared for a sudden offensive or defensive movement; that lookouts

were properly stationed and vigilance observed, and that officers and crew promptly assembled at their quarters.

Eighth. That order was preserved on board, and orders promptly obeyed by officers and crew up the time of the sinking of the ship. In view of the above facts the court have to express the opinion that no further military proceedings are necessary. J. F. Green, Captain and President.[14]

While Union artillery batteries at the tip of Morris Island continued their slow, continuous bombardment of Charleston, Henry J. Leovy, then on assignment in Marietta, Georgia, had apparently heard "vague dispatches" regarding the *Hunley*'s success against the USS *Housatonic* and immediately sent an inquiry to military headquarters.[15]

Marietta, Georgia, March 5th, 1864. General G. T. Beauregard, Commander at Charleston. Sir: Vague dispatches have reached me with reference to the destruction of the Housatonic, and as I am one of the owners of the Torpedo Boat "H. L. Hunley," and also the Executor of the late Captain Hunley, who sacrificed his life in his plans for the destruction of the enemy's vessels. I am exceedingly anxious to learn whether Lieutenant Dixon accomplished his gallant act with our boat or not, and whether he has escaped. It will be a source of infinite pride to me to learn this, even in the least manner, I have been incidentally of service to the general who has gloriously preserved Charleston against the combined attacks of the army and fleet of the enemy. I would therefore beg the favor to be furnished with such information in this matter as General Beauregard may deem it proper to give me. General Beauregard may perhaps remember the writer as one of the proprietors of the New Orleans Delta for several years before the fall of the city. Very Truly and Respectfully, Henry J. Leovy.[16]

In response to Henry Leovy's request (his introducing himself as "one of the owners of the torpedo boat 'H. L. Hunley'" was an intimate connection with the Singer group virtually unknown prior to the discovery of this letter), he received the following informative message without delay: "Head Quarters Department of South Carolina, Georgia and Florida. March 10th, 1864. Sir: I am directed by the commanding General to inform you that it was the torpedo boat 'H. L. Hunley' that destroyed the Federal man of war 'Housatonic' and that Lieutenant Dixon commanded the expedition; but I regret to say that nothing has been heard either of Lieutenant Dixon or the torpedo boat, it is therefore feared that gallant officer and his brave companions have perished. Respectfully H. W. Feilden."[17] Thus, military headquarters seemed well aware that the *Hunley* had never gone to Georgetown, and we can conclude that only the owners of the submarine were made aware that Dixon and his crew were presumed dead.

As news of the sinking of an enemy blockader off Charleston spread across the South, one of the founders of the Confederate Secret Service, B. J. Sage, was then in Shreveport, Louisiana, trying to rally support from military authorities to establish "Bands of Destructionists" (like Captain Singer's group) on the western shore of the Mississippi.[18] Sage was well acquainted with the surviving owners of the *Hunley* submarine, Singer, Dunn, and Braman, for he vigorously encouraged the trio to expand their covert operations into the western theater.[19]

For a little over a month prior to the *Hunley*'s success off Charleston, Sage had met with operatives throughout Louisiana and various high-ranking military officers throughout Gen. E. Kirby Smith's Trans-Mississippi command to discuss the government's views about covert operations against the enemy:

> Headquarters Trans-Mississippi Department, Shreveport, La., December 12, 1863. Major General Richard Taylor. General: The Lieutenant General commanding desires me to commend to your favorable consideration the various measures that may be undertaken to annoy the enemy by organizing small parties, which, acting under your supervision, shall destroy their gunboats, transports and depots. The lieutenant general takes a warm interest in this matter, and would be pleased if, upon investigation, you find it possible to do anything.
>
> He directs me to say Mr. B. J. Sage, a gentleman who will wait upon you, is just from Richmond, and is in possession of the views of the Government on the subject. Mr. Sage has for some time devoted himself to these matters, and can give you information, both as to men and means, which may be valuable in carrying into effect the desires of the commanding general. You can readily appreciate the assistance such parties, properly organized, can render you, especially as the interior rivers and bayous of Louisiana will be soon navigable. Your obedient servant, W. R. Boggs, Brigadier General, and Chief of Staff.[20]

To get some idea what Sage had in mind for General Smith's Trans-Mississippi Department, we need look no further than a lengthy letter (fourteen pages) he sent to General Boggs in early March 1864.[21] After acquainting the general with the names of several skilled individuals in the lower Mississippi region who might be available for covert operations, he moved on to the topic of torpedo warfare and named all Federal vessels that had thus far been sunk by torpedoes (the *Cairo, Baron DeKalb,* and "a large ironclad disabled in the James River").

After further reporting that "the Housatonic was recently destroyed at Charleston," he made the following statements about the *Hunley*, which unbeknown to him at the time, was the vessel responsible for the *Housatonic*'s destruction: "See also the perfect, though fruitless, demonstrations of the

Submarine boat at Charleston. Ought not these cases . . . encourage—nay impel us to efforts here?" Toward the end of his lengthy proposal to General Boggs, Sage makes the following recommendation in regard to deploying Singer torpedoes: "As to torpedoes I should have them worked at proper points where the channel is narrow, using Singer's tried and approved lock."[22]

Sage was well acquainted with various members of Captain Singer's Secret Service Company prior to his deployment to General Smith's Trans-Mississippi Command, because we know that he personally vouched for three of them in a secret communication that eventually fell into the hands of the enemy (the letter of introduction). The capture of this document would cause repercussions from one end of the Mississippi River to the other in the South.

While Sage continued his duties on the western shore of the Mississippi River, Dunn (who had arrived in Mobile from Richmond some four weeks earlier) was then making final preparations to attempt the hazardous crossing himself with Colonels Henry E. Clark and Ward, two officers attached to Missouri guerrilla bands who had been temporarily on duty in Alabama.[23] The three traveling companions were well aware of the dangers that awaited them, for Robert Dunn clearly stated in a letter that the planned attempt was to be made only under cover of darkness.[24] Little could Dunn and his two partners have guessed that at the same time they were making plans to journey to the western shore, Federal forces were being massed up and down the river for an attack against General Smith's rebel army, an attack that would come to be known as "The Red River Campaign."

During this time news of the *Hunley* having sunk the enemy steamer *Housatonic* first reached Mobile. By early March 1864, all that the Singer group presumably knew about their vessel was that it had been responsible for taking the lives of members Gus Whitney, Horace Hunley, and twelve volunteer crewmen, and the commotion the glorious news caused among those then on duty at the Park and Lyons machine shop must have been tremendous.

While Dunn prepared to leave for Texas with the two Missouri cavalry officers,[25] John Braman sat in his hotel room and penned a letter to his wife that included both a summary of the group's intention to build ironclad torpedo boats and a lengthy accounting of his involvement with the *Hunley*. For well over a year Braman and the other founding members of Captain Singer's group had been on duty in the eastern theater, and the following letter given to Robert Dunn prior to his departure from Mobile indicates that Braman was growing somewhat weary of the situation:

Mobile, Alabama, March 3, 1864. My Dear Wife: I write this to send by Bob Dunn, who leaves here Saturday morning for home. I wrote to you when Bob left for Richmond and hinted (for then I did not dare more) at the object of his mission thither, and now, having a more reliable conveyance for my letter, can talk to you more freely of the matter; and, in doing so, will endeavor to explain to your satisfaction why it was mutually agreed between ourselves that Singer and myself should remain on this side of the river and Bob return to Texas.... We at once concentrated all the inventive genius in our party for the purpose of getting up something new that would carry destruction to the Yankees, make money for ourselves, and at the same time be of great service to the Confederacy.

The result was that I got up the plan of an ironclad torpedo boat that, all who saw it admitted, was equal to the task of destroying any war ship now afloat. To carry out our plan and get our new boat under headway, it was necessary, first, to lay the whole matter before the authorities at Richmond, get their approval of the scheme, and authority, together with money and material, to build her.

Secondly, ... the boat would have to be constructed, manned, and used.... We deemed it best, after consultation, to send Dunn to Richmond, and through the influence of Wigfall and others to get the matter before Congress, and authority procured from it to carry out our plans. This part of the business Dunn was well suited for, and has succeeded in accomplishing what we desired.... Next came Singer's duties and my own, which were to superintend the construction and management of the boats after they were built. This requires considerable ingenuity and knowledge of machinery, and for this part of the work no one was at hand but Singer and myself.

This was the program agreed upon, and as neither Singer or myself could go home, and feeling it was necessary someone should be there to take care and look after our families, we agreed that after Dunn had finished his part of the work he should go home.... You can judge for yourself the wisdom of the agreement. That I am more than anxious to go home I hope you have not the slightest doubt; in fact, know that you have not, but you will readily perceive that under the circumstances I must remain on this side for some time, or give up our present scheme entirely.

Bradbury would be of great assistance to us on this side in building our boats, and it is possible that he may come over when Dunn gets home. If so, and you think it at all practicable to come with him, I would like for you to do so. I have talked to Dunn about this, and when you see him, he will be able to tell you all about the trip and counsel you in making the journey.... We are looking daily for news of Singer (who is still in Richmond), after which I shall be able to write you definitely in regard to our future movements. If there is any slip at Richmond in issuing our orders to our entire satisfaction, then all of us will go across the river, so that now I am in a state of happy indifference. If everything is fixed up at Richmond to our satisfaction, it is well; if not, we go home, which is better.

Since we have been on this side of the river we have gotten up a great many projects and have been interested in many new schemes, the particulars of which are too lengthy for an ordinary letter. Among the number, however, was a submarine boat, built at this place, of which Whitney and myself bought one-fifth for $3,000. We took her to Charleston, for the purpose of operating there, and a few days after her arrival there, she sunk through carelessness and her crew of five men drowned.

Another crew of eight men went on from here, raised her, and while experimenting with her in the harbor, sunk her and all eight were drowned. Lieutenant Dixon then went on from here and got another crew in Charleston. A few nights ago he went out, attacked and sunk the steam sloop of war "Housatonic," but . . . fear that he and his crew were all lost. I enclose you a slip from our paper, giving an account of the affair, which will be interesting to you, as Singer and myself built the torpedoes with which the ship was destroyed, and besides we own a considerable interest in the value of the ship, as the owners and crew of the boat got one-half of her value for destroying her. Besides this, we sunk one boat and seriously injured another in the York River in December. . . .

Bob leaves in the morning, and as my sheet is near full and no news to tell, I will close. Write to me at every opportunity and I will do the same. Give my love to all, Mother, Sis, the children and mamma and the boys. I long to see you all and particularly yourself and my dear little Nelly. Kiss her often for her papa and tell her she must not forget him but think of him often and kiss mama for him. I send you a small photograph of myself and intend sending one for mother and one for Sis, but did not have time today to have them taken. With much love I remain Your affectionate and devoted Husband.
J. D. Braman.[26]

Although the letter discusses several pieces of information vital to anyone seeking insight into the shadowy history of the Singer organization, consider the regrettable "particulars" that could have been learned had Braman not considered discussion of the group's "great many projects" to be "too lengthy for an ordinary letter."

Robert Dunn, who set out from Mobile on the morning of March 8, 1864, was given the task of delivering the letter to Braman's wife upon his arrival in Port Lavaca.[27] Some forty-eight hours later, John Braman, on behalf of the Singer Secret Service Corps, received an additional payment of fifteen thousand dollars from the Confederate government for seventy-five recently delivered Singer-Fretwell torpedoes.[28]

Since the Mississippi River and both shorelines were then in the possession of Federal forces, Dunn and his traveling companions may well have been in contact with Confederate operatives in Louisiana assigned the task of ferrying southern military personnel (presumably unfamiliar with the region) to the far shore. From documentation then in Dunn's possession, we know that B. J. Sage had provided him with the names and locations of agents throughout the region, and the three travelers most likely used the safe houses provided by these individuals.

Within a week after their departure from Mobile, the three men found themselves gazing over the moonlit waters of their westward journey's most formidable barrier. For perhaps a night or two, prior to their attempt to cross, they squatted in the rushes at the river's edge and monitored the movements of the Federal patrol boats as they slowly scanned the black water with bright calcium searchlights for any sign of rebels attempting to cross the wide river.

After securing a small boat from unknown sources, the three adventurers set out to cross the enemy-controlled Mississippi on the night of March 16, a decision that would nearly cost the men their lives. Robert Dunn informs us of the near-fatal events that transpired during their attempt to cross the Mississippi: "In crossing the Mississippi River on the night of the 16th instant, in company with Col. Ward and Col. Clark, I had the misfortune to lose my papers, having been closely pursued by launches from a gunboat, and fired at three times from a small swivel gun on their bows, before nearing the shore and twice afterwards."[29]

From Dunn's own account of the three men's narrow escape, the trio hastily abandoned their borrowed boat upon first touching the far shore and quickly scurried into the dark woods, leaving their lost "papers" floating behind them in the muddy waters of the Mississippi. Safely on the opposite shore, most likely they let out joyous whoops and taunting rebel yells directed toward their unsuccessful attackers as the lucky trio dashed ever westward through moonlit groves of trees choked with Spanish moss. Little could they have guessed that the papers they had "lost" (they could well have been thrown overboard when it looked as though the trio was about to be captured) would soon put Federal detectives to work rounding up rebel agents throughout the region.

Among the many documents in their possession was a letter to Colonel Clark from a Confederate torpedo designer, Thomas E. Courtenay, an acquaintance of Captain Singer and leader of Courtenay Secret Service Corps.[30] The Courtenay letter is extremely important to the history of the Confederate Secret Service, for it detailed for Colonel Clark the successful testing and manufacture of Courtenay's "coal torpedo" (a deadly explosive device resembling an ordinary lump of coal that was truly worthy of the name "infernal machine") he had completed and was then ready for distribution to agents throughout the South.

In his letter to Colonel Clark, Courtenay stated that he planned to deploy agents equipped with these devices throughout federally controlled territories for the sole purpose of planting them in the enemy's coal bunkers.[31]

When the innocent-looking lump of coal had been shoveled onto boiler fires in an enemy factory or ship, the evidence, as well as the intended target, would be destroyed in a combination of flame and scalding hot water blown from pressurized boilers.

In addition to the Courtenay letter, Dunn himself had been carrying letters from both Edgar Singer and John Braman that were to be delivered upon his arrival in Texas, as well as his military orders, and list of cooperative southern operatives who might be of service along the Mississippi River. All these papers were recovered by the Federal gunboat *Signal*, the same vessel that according to Dunn had fired on him and his companions "three times from a small swivel gun."[32] The Federals who found the papers sent a report to the secretary of the navy and mistook Dunn and his associates for unlucky Confederate mail couriers who had mistimed their nocturnal crossing:

> Mississippi Squadron, Flagship Black Hawk, Alexandria, La., March 20, 1864. Sir: I have the honor to enclose you some rebel correspondence which was captured by the gunboat Signal a day or two since, while the rebel mail courier was crossing the river. It gives a complete history of the rebel torpedoes, the machine that blew up the Housatonic [as revealed in the Braman letter], and the manner in which it was done. They have just appointed a torpedo corps (I send one of the commissions) for the purpose of blowing up property of all kind. Amongst other devilish inventions is a torpedo resembling a lump of coal, to be placed in coal piles and amongst the coal put on board vessels. The names of the parties are all named in the correspondence, and I send a photograph of one of them [undoubtedly the photograph John Braman had enclosed of himself for his wife and daughter], which, if multiplied and put in the hands of detectives, may be of service. I have given orders to commanders of vessels not to be very particular about the treatment of any of these desperadoes if caught—only summary punishment will be effective. I trust that we will be prepared to avoid any of these machines. I have the honor to be, very respectfully, your obedient servant, David D. Porter, Rear Admiral.[33]

Among the first documents enclosed with Porter's report to the secretary of the navy were Robert Dunn's orders from the Confederate Engineering Department that directed him to report to Gen. E. Kirby Smith: "R. W. Dunn, having been selected for special service, is authorized by the Secretary of War to proceed to the headquarters of Lieutenant-General E. Kirby Smith, commanding Trans-Mississippi Department, to be attached to one of the companies of engineer troops now being organized in that department under the act of congress 'to provide and organize engineer troops to serve during the war' approved 20th March, 1863. A. L. Rives."[34]

Other documents recovered were the February 8, 1864, letter from Edgar Singer to his wife and the Courtenay letter to Colonel Clark. Also captured was the July 14, 1863, engineering report regarding the usefulness of Singer torpedoes and a Confederate engineering order that named all the founding members of Singer's organization. With these sensitive documents then in the hands of the Federals, Admiral Porter issued the following directive on the same day he sent his report to Washington:

> U.S. Mississippi Squadron, Flagship Black Hawk, Alexandria, La., March 20, 1864. General Order, Number 184. The enemy have adopted new inventions to destroy human life and vessels in the shape of torpedoes, and an article resembling coal, which is to be placed in our coal piles for the purpose of blowing the vessels up, or injuring them. Officers will have to be careful in overlooking coal barges. Guards will be placed over them at all times, and anyone found attempting to place any of these things amongst the coal will be shot on the spot.
>
> The same policy will be adopted towards those persons who are caught planting torpedoes, or floating them down, or with any of these inventions in their possession. Extra vigilance will be required in preventing the passing of boats across the different rivers. . . . The names of the persons who are engaged in the torpedo business are: R. W. Dunn, E. C. Singer, J. D. Braman, J. R. Fretwell, C. E. Frary, F. M. Tucker, L. C. Hirshberger, and the sooner they are got rid of the better. David D. Porter, Rear-Admiral, Commanding Mississippi Squadron.[35]

Unquestionably the greatest coup Admiral Porter and the officers attached to his Mississippi Squadron scored was the recovery of Robert Dunn's letter of introduction given to him by the founder of the Confederate Secret Service, B. J. Sage. This document turned out to be invaluable to the Union war effort because it listed the names and locations of nearly fifty Confederate agents and saboteurs then operating along the Mississippi River. Admiral Porter sent a copy to the secretary of the navy and issued a directive to arrest all those named: "Secretary of the Navy Gideon Wells. Sir: I have the honor to enclose you an order I had printed for the information of the provost marshals, detectives, and others, to enable them to arrest the parties engaged in this torpedo business. The letter printed at the end of the order is one captured in the rebel mail, and was to be used as a letter of introduction by R. W. Dunn, E. C. Singer, and others, to help them along in their scheme. David D. Porter, Rear Admiral, Commanding Mississippi Squadron."[36]

Dunn's letter of introduction was copied word for word so the Federal officers could familiarize themselves with the names of the various rebel agents:

> To introduce R. W. Dunn, E. C. Singer and J. D. Braman to my friends:
> B. C. Adams, Grenada; Captain Samuel Applegate, Winona; Colonel

H. H. Miller, commanding regiment east of Granada and Carrollton; W. P. Mellen, Natchez; Major John P. Peyton, Raymond; Judge D. H. Prosser, F. A. Boyle, Woodville; Henry Skipwith, Clinton, La.; Conrad McRae, Fordoche, La.; W. Barton, J. J. Morgan, T. G. Calvit, James E. Lindsey, Wm. M. Lindsey, Wm. H. Neilson, Samuel Faulkner, Atchafalaya River, La.; Colonel James M. Porter, Colonel Wm. D. Davis, Colonel Wm. Offatt, Captain James Capps, S. A. Scribner, Elbert Goull, T. C. Anderson, Simon Richards, St. Landry, La.; Henderson Taylor, S. L. Taylor, Marksville, La.; H. Robertson, S. W. Henarie, Governor T. O. Moore, Colonel C. Manning, Alexandria, La.; General M. Wells, General P. F. Kearny, Hugh M. Kearny, esq., Lafayette Parish, La.; Hon. John Moore, Wm. Robertson, St. Martins Parish, La.; Judge Baker, T. J. Foster, Judge Palfrey, St. Mary's Parish, La.; Daniel Dennet, editor "Planter's Banner"; Mr. Sickles (Kindred Spirits), Phanor Prudhomme, esq., St. Mary's Parish; John Blair Smith, Natchitoches Parish, La.; Colonel H. J. G. Battle, Rueben White, Caddo, La. We must all help one another, and those who can be efficient in our cause must receive all necessary hospitality, aid, and information. I introduce none but the worthy. B. J. Sage.[37]

Many Civil War historians consider capture of these documents to have been the greatest wartime blunder of the Confederate Secret Service, and Dunn and his two traveling companions probably never knew that these letters of introduction fell into enemy hands.[38] If the letter concerning detailed information regarding Captain Courtenay's secret coal torpedo had not fallen into enemy hands, the Union may have never discovered that southern saboteurs were using this explosive device.

Since Singer operative John Braman's photograph was then being copied and given to detectives, his life would be placed in great peril if he ever attempted the hazardous crossing through enemy-controlled territory. With the names of the core members of Captain Singer's group then in the possession of Federal authorities, the war took on a new complexion for those middle-aged Masons who had accompanied fellow lodge members Singer and Fretwell on their eastward adventure, for they would be shot immediately if captured and identified as a member of the Singer organization.

As Dunn and the two cavalry colonels made their way westward for the Texas border, the combined Union force committed to the Red River Campaign continued its advance toward Shreveport. Federal scouting parties and patrols were searching the region for signs of the enemy, so Dunn and his traveling companions, then presumably devoid of their papers, would have moved cautiously through the area while continually on the look out for friendly forces. We do not know how the group linked up with Confederate forces, but Robert Dunn arrived in Houston some three weeks later.

Since Dunn had no orders or documents then in his possession, he penned the following letter of introduction to General Magruder that outlined the various activities undertaken by "E. C. Singer and Company":

> Houston. April 6th, 1864, Maj. Gen. J. B. Magruder, Commanding The District of Texas, New Mexico and Arizona, C. S. A. General: We beg leave very respectfully to report that since your permission to report to Gen. E. Kirby Smith, at Shreveport, La. In May last, for torpedo service, that we were permitted by him, to report to the Secretary of War at Richmond, Va. Which we did in July last. Upon presenting Certificates of the Engineers of your Department, as also others from Generals Buckner and Ledbetter, of tests made in all the Departments named, it was determined by the Secretary of War to give us every possible facility for using our torpedoes.
>
> We were at once transferred by him to the Engineer Troops and ordered to report to General Joseph E. Johnston at Morton, Mississippi. Leaving operators at Richmond, Wilmington, Savannah, Mobile and Charleston. While operating in the later place (Charleston) it was fully demonstrated, that Torpedoes could be used on the prow, placed on the bow of a boat, without any damage to herself, yet carrying certain destruction to the vessel attacked.
>
> On the first of November last, the writer returned again to Richmond, to obtain assistance for the construction of such boats as would enable us to operate with safety to our crews, and at the same time, strike terror to the enemy. After considerable hesitation, we were finally ordered by the Secretary of War, to construct one boat at Selma, Alabama, and one at Wilmington, N.C., of the following dimensions vis. 160 feet long, 28 foot beam and 11 foot hold with flat deck, carrying all their machinery below—to be iron sheathed and with no capacity for guns, and only showing 2 feet above water when ready for work. They [are] to be arranged with torpedoes, worked from below decks, and through tubes, forward, Aft, and on both sides. It is believed by Engineers of the highest rank, after a full investigation of our plans, that these boats will be perfectly able to raise the Blockade of all the Harbors in the Confederacy.
>
> Quite a number of small boats, known as "cigar boats," for night attacks are now being constructed at Richmond, Wilmington, Charleston and Mobile. Our success in the use of the stationary Torpedo, has been the destruction of the Enemy Transport Gray Cloud in the Atchaffalger Bay, La., in June last, killing and drowning 140 men; The Gun boat Dekalb, of 13 guns on the Yazoo River in July last destroying her entirely with 180 men; A gunboat in York River (name not recalled) was blown up by one of them, destroying the boat and crew and injuring another Boat badly that lay near by.
>
> The sloop of war Housatonic was destroyed in Charleston Harbor, by one of our torpedoes, attached to the prow of a small submarine boat, propelled by nine men. In Tennessee we have blown up eight Rail Road Trains. These successes show conclusively the certainty of explosion of our torpedoes and

we would add, that with the use of the safety line, instead of the trigger, we could work such boats as we are building over a bar where it lacked 4 or 5 feet of water to let them pass over, as also through rafts and other obstructions.

In Feb. last the writer was ordered to report to Lt. Gen. E. K. Smith of the Trans Miss. Dept., for torpedo service, with instructions to him to furnish us with such assistance as he might think the service required in his Dept. In crossing the Mississippi River on the night of the 16th instant, in company with Col. Ward and Col. Clark, I had the misfortune to lose my papers, having been closely pursued by launches from a gunboat, and fired at three times from a small swivel gun on their bows, before nearing the shore and twice afterwards.

My assistance being at Houston and Lavaca, I was induced to come direct to Houston, but have reported to Gen. Smith by letter, and within a few days lay before you our propositions for forcing the Enemy's Fleet Lying off our Bars and Harbors, to take some of our medicine. Permit us General to Express to you our sincere thanks, for permitting assistance given us in developing the utility of our torpedoes in the Department which placed us at once and prominently before the War Department at Richmond, and secured the success above alluded to. Very Respectfully, your obedient servant R. W. Dunn for E. C. Singer & Co.[39]

While Robert Dunn was arranging to meet with General Magruder, Adm. David Porter and several ironclads attached to his Mississippi Squadron were then slowly steaming up the Red River toward the Confederate Naval Yards at Shreveport. In an attempt to stop just such an advance, Gen. Kirby Smith, some weeks earlier, had ordered Capt. David Bradbury and several operatives attached to Singer's Port Lavaca torpedo facility north to Shreveport with as many torpedoes as were then available. Upon their arrival in northern Louisiana, Bradbury's Texans were ordered to anchor the torpedoes (about thirty in number) several miles below the city.[40] On April 15, 1864, David Bradbury's efforts paid off when a detonating rod on one of his submerged torpedoes was jarred loose by the passing hull of the Federal ironclad USS *Eastport*.

Capt. Thomas O. Selfridge (commander of the USS *Cairo* when it was sunk in the Yazoo River) wrote an account after the war regarding the first confirmed Singer torpedo strike to have taken place west of the Mississippi River:

> The Eastport (Lt. Commander Phelps), the largest of our iron-clads which had joined the squadron for the first time on this expedition, unfortunately struck a torpedo eight miles below Grand Encore, and her bottom was so badly injured that she sank. Captain Phelps was very proud of his ship, and went to work with a will to save her. After the most untiring efforts he succeeded in bulkheading the leak, and, assisted by two steam-pump boats which the admiral had brought to his assistance, succeeded in getting her some forty miles down the

river. Here she grounded again, but after strenuous efforts, assisted by the admiral who remained behind, she was floated, but after proceeding a few miles again grounded on a pile of snags. From the 21st to the 25th of April Captain Phelps, one of the bravest and most competent commanders in the squadron, had worked day and night with his officers and crew to save his ship, but the retreat of the army had left the banks of the river unprotected and the low stage of the water had compelled the admiral to send his squadron to Alexandria. There was no longer a chance to save the Eastport, and he reluctantly gave the order to blow her up.[41]

Gen. Kirby Smith's headquarters in Shreveport was well aware that Dunn was back in Texas, for within hours of hearing reports that the USS *Eastport* had been sunk, General Smith sent the following dispatch: "Shreveport, La., April 15, 1864. Major General Magruder, Houston, Texas. Sir: General Smith desires that R. W. Dunn report here immediately with his torpedoes."[42] Upon receipt of this communication, Magruder immediately ordered Dunn eastward with an additional directive to the Houston quartermaster stating that Dunn and his assistants should be furnished with "transportation by rail and stage."[43]

During the weeks that followed, several torpedo boats of varying sizes were ordered constructed in different locations throughout General Smith's Trans-Mississippi Department. And since Captain Singer and his associates were by this time accomplished submariners, it is not surprising that at least some of the torpedo vessels General Smith ordered were submarines.[44] Evidence that Captain Singer's organization would soon again be dabbling in submarine construction comes from a confidential report filed by a Union spy who had been briefed about the project by a talkative Singer agent whom the spy had met while traveling aboard a stagecoach across the Texas prairie. He reported the following regarding five Confederate torpedo boats that appear to have had an uncanny resemblance to the then lost *H. L. Hunley*:

> The following is a description of the torpedo boats, one of which is at Houston and four at Shreveport: The boat is 40 feet long, 48 inches deep, and forty inches wide, built entirely of iron, and shaped similar to a steam boiler.... The boat is usually worked 7 feet underwater and has four dead-lights for the purpose of steering and taking observations.... The air arrangements are so constructed as to retain sufficient air for four men at work and four men idle for two or three hours. The originator and constructor of these boats, also constructed the one which attempted to destroy the New Ironsides in Charleston.[45]

While negotiations for the construction of various Singer-designed torpedo boats were taking place at General Smith's headquarters in Shreveport, Edgar Singer himself was still attending meetings in Richmond. By the early spring of 1864 the Confederate Congress had passed a bill that modified the recruitment of agents for missions behind enemy lines, as well as the status of individuals then attached to groups such as Captain Singer's. The congressional act sanctioned the creation of sabotage organizations similar to Singer's group; however, the new teams of agents were not in future to be unattached to the Engineering Department.[46]

Singer's well-established team of torpedo engineers was soon to be placed under the jurisdiction of this new branch of the War Department, and apparently official transfers for Singer's core members would have to be made before reestablishing his previously organized Secret Service Company:

> Richmond, April 28, 1864. Hon. James A. Seddon, Secretary of War. Sir: J. D. Braman, R. W. Dunn, C. E. Frary, James Jones and David Bradbury together with myself were by order detached from Shea's Battalion of Artillery (Texas) to be attached, as you may remember, to the Engineer troops for the purpose of affording us protection in the operation of the torpedoes of my invention against the enemy. At that time this was the only plan which could be devised to place us upon the footing of prisoners of war in case of capture.
>
> This was done with the expectation and understanding that Congress would soon pass an act giving special organization to this service into which we were to be incorporated, and accordingly on the 21st of March 1864, you were pleased under the act of congress to grant me authority to raise twenty five men for this Special Service. The parties named have all along been connected with me and participated in what we have accomplished.
>
> Their energy and skill are essential to my corps. But in order to carry out the purposes which you assigned to us, it will be necessary to have an order from you authorizing transfers of these men from Shea's battalion to my company. Time now is everything to us, and I have to ask in accordance with the duty signified to us that you grant me an endorsement directing this transfer. I shall immediately proceed to the execution of our plan of operations which will inflict punishment upon the invading enemy. Respectfully your obedient servant E. C. Singer.[47]

A tattered report filed by Capt. J. Andrews (officer in charge of torpedo deployment in Mobile Bay) with the Confederate Inspector General's Office some months later confirms that Singer's requests were granted. He included the "List of Detailed and Employed Men on duty in Torpedo Service District of the Gulf, Mobile Alabama." Of the diverse groups then occupied in various aspects of torpedo manufacture, deployment, and maintenance, only those individuals listed as being attached to "Captain McDaniel's Company of Secret Service" and "Singer's Secret Service Corps" are designated as being under the direction of the War Department (under the column

"To whom Detailed," other individuals and groups have notations such as Conscript Bureau, Enrolling Officer, General Maury, General Johnston, and General Gardner).[48]

Of the men listed as members of Captain Singer's Secret Service Corps then on duty in Mobile, James R. McClintock, Baxter Watson, John A. King, and Greenleaf Andrews, only the backgrounds of McClintock and Watson are known to any degree, so we do not know how the organization acquired Andrews and King. It is also surprising why none of the names of the founding members appear on this document, because Mobile seems to have been the hub of the group's operations. It may be that these founding members were involved in operations such as construction of an ironclad torpedo boat in Wilmington and Selma. The circumstances surrounding the assumed whereabouts of these Texans are examined in depth later, but we here follow the early-spring redeployment of John R. Fretwell to the Confederate capital at Richmond.

Within days after Captain Singer had filed his April 28, 1864, request with the secretary of war for transfers to be issued to several members of his group, Singer himself seems to have returned to Mobile and turned the Richmond operation over to his top lieutenant, Fretwell. We know that Fretwell was indeed in Virginia by early May from documentation recently uncovered at the National Archives.[49] Just why Fretwell journeyed to Richmond while Singer returned to Mobile may remain unknown, for he seems to have knowingly walked right into the Federal army's spring offensive to take Richmond.

In late April 1864, the latest of Lincoln's commanding generals, Ulysses S. Grant, was preparing to launch a two-pronged campaign to capture the Confederate capital. While his Army of the Potomac pushed southward from the Rapidan River, Gen. Benjamin Butler would simultaneously ascend the James River accompanied by numerous gunboats and warships attached to Adm. S. P. Lee's North Atlantic Blockading Squadron.[50]

John Fretwell arrived in Richmond during this time of high drama. Whether Captain Singer was still in Richmond is unknown because the only relevant torpedo-related documents directed toward his group are addressed to Fretwell.[51] With Federal gunboats then steaming up the James River and General Lee's army desperately trying to stop Grant's advance southward, Gen. Gabriel Rains was immediately summoned back to Richmond from an inspection tour he was conducting in Mobile.[52] Whether Fretwell hastily accompanied Rains and his small staff back to Richmond is not known, for we have only a few disjointed dispatches placing him in the capital under General Rains's command during the month of May.[53]

When General Rains arrived in Richmond, he changed torpedo tactics in the James River to those first introduced by the American patriot David Bushnell against the British fleet at Philadelphia during the Revolutionary

War ("The Battle of the Kegs").⁵⁴ With enemy ships slowly steaming toward the southern capital, Rains found himself in practically the same situation as George Washington during the fall of 1777, so Rains deployed a deadly mode of warfare, various forms of drift mines. In the early-morning hours of May 13 the first Confederate mine was discovered floating down the James River, and the following report was hastily penned and sent up the chain of command:

> USS Pink, Off Bermuda Hundred, May 13, 1864. Sir: I have the honor to report that this morning about 5 o'clock the officer of the deck had his attention directed to a piece of board drifting towards this vessel. It was about two feet long and about one in width, evidently having something attached to it. We threw a small fishing line over it, and held it until we lowered a boat and made a small line fast to it and towed it about thirty feet from this vessel, when it exploded, without injury to this vessel or to the boat. The torpedo was of tin, about fifteen or eighteen inches in diameter and about 2 feet in length, and in shape much like a milk can. Your obedient servant John W. Dicks, Commanding USS Pink.⁵⁵

Within hours another type of drift mine was recovered from the waters of the James, prompting the commander of the USS *Hunchback* to send the following message to Adm. S. P. Lee: "USS Hunchback, Off City Point, James River, Virginia, May 13, 1864. Sir: I have the honor to report to you the success of capturing a large torpedo in the river as it was floating down in a direct line for this vessel. The can contained about 75 pounds of fine rifle powder, which was in a perfectly dry state when the can was opened. I send you the can, together with a sample of powder in it. R. G. Lee, Commanding.⁵⁶

By this time Gen. Benjamin Butler and the numerous Federal regiments under his command had advanced to within some twelve miles of Richmond and were in desperate need of support from the admiral's gunboats. The number of floating mines was assumed to increase dramatically with each additional mile one got closer to Richmond, so the following dispatch could not have been welcome aboard Admiral Lee's flagship: "In The Field, Near Drewry's Bluff, May 13, 1864. Rear-Admiral Lee, Commanding. Sir: Would it not be possible for you to bring up the gunboats, monitors, opposite Dr. Howlett's, so as to cover our flank on the river and relieve a considerable body of my troops? Both sides of the river there are low and flat, and it is an excellent point for the gunboats to lie. Benjamin F. Butler, Major General."⁵⁷

While General Butler's forces continued their advance along the James River shoreline and Admiral Lee's fleet steamed ever closer, John Fretwell and the Singer operatives then assigned to Richmond presumably worked around the clock to rig and deploy as many drift mines as possible. From various dispatches we know that Fretwell and his associates prepared their torpedoes in close proximity to General Butler's advancing army, a duty undoubtedly performed with some apprehension, for all torpedo men knew that immediate execution was the sentence for anyone captured with these devices in their possession.[58]

Gen. Robert E. Lee was well aware of Fretwell's torpedo activities since he had penned a document stating that John Fretwell had personally "received my permission to operate on the James River."[59] Additional documentation compiled after the war claims that the Texas physician had also served for a time on General Lee's staff sometime during the final year of the war.[60] We do not know when this assignment might have taken place (Fretwell served in the vicinity of Richmond for the remainder of the conflict), for Lee had stated that he had personally placed Fretwell "under the direction of General Rains."[61]

By May 1864, General Rains had crossed paths with the Singer group several times and was obviously well acquainted with the group's various members. He had been on duty in Charleston when the *Hunley* was operational and had even been present in General Beauregard's office on the morning that Lieutenants Dixon and Alexander reported the details of their submerged duration test. General Rains may well have regarded the Singer organization to be somewhat extreme in their tactics, but by early May 1864, their unorthodox methods of underwater warfare had proven effective on several fronts, so the general would have undoubtedly welcomed the services of Fretwell and his associates.

On the morning of May 21, some eight days after the first mine had been set adrift in the James River currents, the following urgent telegram was sent from Col. John Williams's headquarters several miles south of Richmond: "Dr. J. R. Fretwell: Can you raise some men and bring down the eight torpedoes via Drewry's Bluff [stop] I can detail a few men here [stop] I want them assigned for floating down upon the monitors that are shelling us. . . . McDaniel has disappointed me."[62]

The reason that Capt. Zedekiah McDaniel disappointed Colonel Williams might have been that his group had been assigned the task of mining General Butler's assumed route toward Richmond with "subterra shells,"[63] a contact-sensitive land mine designed to explode from the weight of a man's foot. Although several types of drift mines were employed against enemy vessels in the James River during the opening weeks of May, none found their mark, so they probably produced little more than a psychological effect on the enemy.

By the closing days of May it was painfully clear to the Confederates that this newest of Lincoln's generals, Ulysses S. Grant, was not going to return northward after his army had been bloodied in the field (as they had done in past years). Instead, Grant rallied his forces after being surprised and pummeled at the Battle of the Wilderness and slowly continued to move ever southward toward his objective.[64]

With overwhelming forces at his command, Grant, whose army outnumbered Lee's nearly two to one, knew it was only a matter of time before the badly depleted ranks of General Lee's Army of Northern Virginia would be pounded into submission. This willingness to trade high casualties with Lee's army would lead many in the North to openly call Grant a butcher who considered the men under his command to be little more than cannon fodder. In spite of the appalling casualties being amassed, Lincoln continued to endorse Grant's actions and openly shared the general's view that the South would not abandon its struggle for independence until its armies were crushed and they realized that further resistance was useless.[65]

While the fields of Virginia were once again being stained with American blood, Singer operative R. W. Dunn, who had been summoned to Shreveport by Gen. Kirby Smith some weeks earlier, continued to meet with the general's engineering officers regarding the construction of torpedo boats west of the Mississippi. Four small torpedo boats of the Singer design were to be immediately constructed at Shreveport for the defense of the Red River, and two huge ironclad vessels would be built at Houston for use against the Federal blockading squadron off Galveston.[66] From the following order issued by General Smith, we can get some idea of the importance the general attached to the construction of these huge offensive weapons slated for deployment along the Texas coast:

> Headquarters Trans-Mississippi Department, Shreveport, La. May 19, 1864. To General Magruder Commanding the District of Texas. Special Order 123. Mr. R. W. Dunn Engineering Troops, will proceed to construct under the direction of the Chief Engineer of the Trans-Mississippi Department, two torpedo boats of the dimensions proposed in his communication of 21st of April.
>
> Mr. D. Bradbury is hereby appointed Chief Constructor and Jas. E. Haviland Assistant Constructor who will report to and be guided by the instruction of Mr. Dunn. The boats will be constructed on Buffalo Bayou at or near Houston. Such machinery, iron etc. as may be required in the construction of the boats will be procured by Mr. Dunn. The Commanding General of the District of Texas will render such assistance to Mr. Dunn as may be necessary in procuring material and mechanics for the construction of the boats. By Command of General E. Kirby Smith.[67]

Robert Dunn's letter of introduction to General Magruder described the unique weapons system and vessel that General Smith ordered to be constructed on Buffalo Bayou. From interviews conducted several months later with Confederate deserters familiar with the project, we know that the huge torpedo boats appeared rectangular in shape and were sheathed in railroad iron.[68] They did not reveal how these vessels were to be "arranged with torpedoes, worked from below decks, and through tubes, forward, aft, and on both sides,"[69] nor did they reveal how this revolutionary new form of underwater weaponry was to be utilized.

We do not know how designers Robert Dunn and John Braman first conceived of the idea to incorporate self-propelled torpedoes as the primary weapons system for their proposed ironclad torpedo boat. However, experimentation with just such a rocket-powered device took place in the Mobile River at exactly the same time that the *Hunley* was under construction at the Park and Lyons machine shop.[70]

General Smith apparently considered underwater torpedoes an effective form of weaponry, so he sent orders to Houston headquarters to construct the boats (a Confederate document recently discovered at the National Archives states that some form of rocket factory was located at Galveston, Texas).[71] On the day after General Smith had ordered General Magruder to assist Dunn in any way necessary, Magruder's headquarters sent the following request: "Houston, May 20, 1864. Special Order 141. Private I. R. Hutcheson 8th Texas Cavalry will report for extra duty on Torpedo Service to R. W. Dunn esq. At Shreveport, Louisiana without delay. The Quartermaster will furnish transportation by stage and rail."[72] Robert Dunn was well acquainted with Ike Hutcheson (a fellow citizen of Port Lavaca) and had some confidence in his administrative skills, for within several months, Hutcheson took over some of Dunn's responsibilities and was placed in charge of all torpedo operations on the Red River.[73] Private Hutcheson may well have been working with the Singer group at Port Lavaca prior to his redeployment to Louisiana, because he may have been redeployed to Shreveport to free up Dunn for other torpedo-related duties, which may have had something to do with the following order issued while Hutcheson was presumably en route to Shreveport:

> Headquarters Engineering Department, Trans-Mississippi Department, Shreveport, La. May 25, 1864. Mr. R. W. Dunn Torpedo Service. Sir: You will proceed to Houston, Texas and make arrangements without delay for constructing two torpedo boats in Buffalo Bayou. You will procure with the aid of the commanding general of the District of Texas such materials that may be necessary, reporting cost, etc. to this office.
>
> You will be supplied with funds from time to time, as may become necessary—the disbursements will be made by an officer whom I shall appoint. Reports of progress made with return of Assets and Hind men will be made by you regularly at the end of every month, to this officer. You will make your

returns in accordance with forms one and two Army Regulations for the Guidance of Officers of the Corps of Engineers. Very Respectfully, Your Obedient Servant H. I. Douglas, Major and Chief Engineer."[74]

With Robert Dunn's new duties may well have come an upgrade in rank. Among documents under "Miscellaneous Torpedoes" in the Confederate Navy Subject File at the National Archives is a small, tattered envelope addressed to "Captain R. W. Dunn, Shreveport, La." In an order issued from Houston headquarters less than two months later, he was addressed as "Chief of Torpedo Service."[75] He, like Edgar Singer, David Bradbury, and the late Horace Hunley, had by early 1864 been elevated to the rank of captain in the Confederate military.[76]

Robert Dunn was induced to temporarily shelve his torpedo boat diagrams within days of his arrival in Houston in an order issued from Magruder's headquarters: "Houston, June 6, 1864. Special Order 158. Captain H. B. Lee Assistant Quartermaster at Beaumont will furnish transportation to Mr. R. W. Dunn, and one man, and torpedoes to Calcas, Louisiana."[77] (Calcasieu Parish, located in the heart of Louisiana's French-speaking Cajun country, borders the Sabine River that separates Texas from Louisiana.)

Before leaving for Louisiana, Dunn apparently discovered that he and his assistant would require additional gunpowder to prime their mines, because Magruder's headquarters sent the following orders: "Houston, June 7, 1864. Special Order 159. Captain Henry T. Scott will turn over to Mr. R. W. Dunn of the Engineering Department one hundred and twenty five (125) pounds of good powder to torpedo purposes. The Quartermaster will furnish transportation for same to Niblets Bluff."[78] Niblett's Bluff was the Confederate staging area for operations throughout western Louisiana and along the Sabine River,[79] and Dunn's mission to the region most likely was to obstruct the vulnerable waterway with Singer torpedoes.[80]

While Robert Dunn and his associates reported for duty at Niblett's Bluff (and presumably searched for suitable locations to deploy their torpedoes in the stagnant, alligator-infested waters of the Sabine River), Singer operative David Bradbury, who had been appointed "Chief Constructor" of the Houston torpedo boat project on May 19, sent the following report to Houston headquarters:

> Houston, June 15, 1864. To Captain J. S. Aldrich, Capt., A.A.G. Sir: Acting under Special Order number 123, of Lieutenant General E. Kirby Smith, dated at Shreveport May 19, 1864, I have verbally contracted with Mr. H. Close and Sons of Galveston, to make such boilers and machinery as is necessary for the construction of the two Torpedo Boats named in said order, but before proceeding with said work they wish an order to be issued from Headquarters, ordering them to do said work, and so worded as to protect them from being

interfered with by any other branch of the government while in process of construction. Very Respectfully, D. Bradbury, Chief Constructor.[81]

From Captain Bradbury's report it seems that Close and Sons were somewhat overwhelmed by the complexities of the project and decided then and there that the undertaking would require the undivided attention of all those then employed at their facility. Since contemporary documents record that the Confederate government furnished locomotive steam engines for the Singer-designed torpedo boats,[82] the machinery to be fabricated by Close and Sons likely had nothing to do with the propulsion system. The boats were designed to employ a revolutionary new weapons system, so the proposed fabrication of these devices may have prompted Close and his machinists to make the requests. Robert Dunn immediately upon his return to Texas sent a similar request to Houston military headquarters, as summarized in a book of endorsements: "R. W. Dunn Requests that an order be issued that Mr. Close shall not be molested while building machinery for the two torpedo boats."[83]

Before Dunn returned to Texas, he discovered (or was perhaps directed to) a large quantity of iron aboard a steamer anchored on Lake Charles (in the center of Calcasieu Parish). Either he requested that the iron be turned over to him or military authorities directed him to haul it back to Houston for his torpedo boats: "Houston, June 21, 1864. Special Order 173. Captain H. Lubbock will have the iron taken off the steamer 'Wave' at Lake Charles, and turned over to R. W. Dunn for torpedo service, and report the audit to these headquarters."[84] (Capt. Henry Lubbock was the Texas governor's brother and commander of the Marine Department for the District of Texas.)[85]

Dunn was in Houston by July 2, and whatever other duties he was performing during this period can be summed up in the following directive issued from Houston Headquarters on July 8: "Houston, Special Order 190. In obedience to instructions from Department Headquarters Trans-Mississippi Department, the Marine Department will furnish necessary transportation for the movement of Negroes, tools, material, etc., engaged in the manufacture of torpedo boats, upon the request of Mr. R. W. Dunn Chief of Torpedo Service, or his authorized agents."[86]

The directive could be interpreted that additional torpedo boats of unknown design were then being constructed at locations other than Houston. And since men "engaged in the manufacture of torpedo boats" seem to be requiring transportation to unspecified locations, apparently other Singer projects had been started by that time. Other evidence pointing to additional "secret" Singer torpedo boat projects is the following order filed by Federal rear admiral David Porter:

Mound City, June 25, 1864. Lieutenant-Commander F. M. Ramsey, Commanding USS Choctaw and 3rd District Mississippi Squadron. Sir: The rebels are fitting out at Shreveport four torpedo boats. They will be ready in two months. You will at once prepare a chain across the mouth of Red River on floats, and so fitted that it can be opened to permit vessels to go in and out. By keeping the chain constantly on the stretch and a small gunboat lying close to it at the mouth of the river guarding it, no torpedo boat can get out. Obtain all the information you can about this and do not be caught napping. Do not forget that Old River is also open at times. Very respectfully, your obedient servant, David D. Porter, Rear-Admiral.[87]

Just where Admiral Porter received this previously unrevealed information is not known, for no Confederate records thus far discovered discuss the construction of torpedo boats on the Red River during the spring of 1864. Although no southern documents have come to light regarding the fabrication of these boats, a report filed by a Union spy several months later includes a report that not only describes the vessels in graphic detail but also gives an outline of their intended mission, revealed to him by a talkative Singer agent.

By late June 1864, both Edgar Singer and John Braman were back in Mobile. Apparently the Confederate War Department had not forgotten about them: "June 24, 1864, To E. C. Singer, Care of Lieutenant-Colonel Von Scheliha, Chief Engineer Mobile, Alabama. Sir: Details for yourself, Braman and others forwarded today by mail. A. L. Rives, Colonel."[88] Orders sent from Richmond by mail on June 24 did indeed send both Singer and Braman back to Texas. In General Magruder's order books Braman is listed as the "superintendent" of construction on one of the Houston torpedo boats.[89]

John Fretwell, then in charge of Singer torpedo operations in central Virginia, continued with his duties on the James River. By early July he had been granted a somewhat free hand in deciding when and where the James River was to be mined. Although General Lee had issued orders placing Fretwell and his assistants under the command of General Rains's Torpedo Bureau sometime during the spring, some confusion had developed between the various personnel engaged in mining waterways that led to Richmond.[90]

In April 1864, with Federal operations against Charleston at a near standstill, General Beauregard and several officers in his command were ordered north to Virginia to assist in defense of the capital. Among them was Capt. Pliny Bryan, an expert in the deployment of the Rains keg torpedo who had recently been responsible for sinking four Federal transport vessels on the St. John's River in northern Florida. Upon Captain Bryan's arrival in Virginia, he was placed under the command of General Rains and given the duty of mining the river with contact-sensitive keg torpedoes. From the following report filed with his old commanding officer in mid-July, we can assume that some friction had developed between himself and various members of the Singer organization:

> Headquarters Department of North Carolina and Southern Virginia. July 18, 1864. General Beauregard. Sir: I have the honor to report that on Thursday evening last I started with a torpedo expedition from Chaffin's farm for the James River. At first I intended to operate on Harrison's Bar, near Berkeley (the place I selected sometime since), But finding Doctor Fretwell had selected the same place and for the same purposes, and being informed by him that he was ready to operate and was acting under the orders of General R. E. Lee, I made a reconnaissance lower down the river and selected Westover.
>
> Everything being ready, the expedition embarked about sunset. . . . About 1 o'clock, and just as the expedition was in the act of leaving the shore, a steamer was heard coming up the river. . . . It soon became evident that the steamer intended making a landing at Westover, which she did and immediately put a force on shore. . . . Several shots were exchanged between the enemy and our men. . . . With the balance of the party, six men, I left making our escape by cautiously following the margin of the marsh. . . . The expedition consisted of two rowboats and twelve torpedoes complete, all of which fell into the enemy's hands. . . . I regret the result exceedingly. E. Pliny Bryan, Captain.[91]

With both his boats and torpedoes then in enemy hands, Captain Bryan seems to have blamed the entire affair on Fretwell's expedition that mined the James River at a point that he (Bryan) had previously selected. Beauregard quickly filed a report on the incident with Gen. R. S. Ewell, who in turn requested clarification from General Lee himself. Lee responded with a letter regarding the duties he had assigned to Fretwell:

> Headquarters Army of Northern Virginia, July 25, 1864. Lieutenant General R. S. Ewell, General: Your letter of the 23rd instant is received. I had understood that General Rains was in charge of the torpedoes on James River, and when Captain Bryan was sent from here I desired him to report to General Rains, or whoever might have the direction of the subject. Dr. Fretwell was brought to me by Lieutenant-Colonel Williams, of the engineers, who stated that Dr. Fretwell was acting under the chief engineer at Richmond. He applied for and received my permission to operate on James River so far as that permission was necessary, but I understood that the whole matter would be under the control and direction of General Rains, to whom he was directed to report. I did not contemplate sending independent parties. If General Rains is not in charge you had better select some competent officer, and put the whole matter and all the persons engaged in it under his control. Very respectfully, your obedient servant, R. E. Lee, General.[92]

As further clarification, General Lee sent a note to General Beauregard, who was then commanding ragged troops in the muddy trenches south of Petersburg: "Doctor Fretwell is acting under no special orders from me. He was directed to report to General Rains."[93] In light of the mishaps that caused such repercussions throughout the Confederate high command, General Rains and the officers assigned to his Torpedo Bureau most likely closely scrutinized future torpedo activities along the James River.

As strange as it may sound to the student of Civil War history, the summer of 1864 was by no means such a dismal time for the Confederacy. Although the Battle of Gettysburg had been fought some twelve months earlier (considered by many historians to have been the turning point in the conflict), the Confederate flag still flew defiantly over Richmond, while Ulysses S. Grant's battle-weary troops were stuck in muddy trenches south of Petersburg. Although Grant had continually moved southward, he had never been able to deliver the all-important death blow to Lee's Army of Northern Virginia and had himself suffered appalling casualties.[94]

To the south the Army of Tennessee stood between General Sherman and Atlanta, and Gen. Jubal Early's advance into Maryland, which culminated at the Battle of Monocacy, had forced the Federals to pull troops away from Grant to protect Washington. The Union army in the lower Mississippi Valley was still recovering after Gen. Kirby Smith had soundly routed them in the Red River Campaign, and European munitions and supplies still moved through the blockade into the harbor cities of Wilmington, Charleston, and Mobile. The New York draft riots of 1863 and war weariness in the North had caused numerous Federal politicians to rethink the conflict, and several openly voiced their opinions during the summer of 1864 that the time had come to negotiate an armistice with the Confederacy.[95] Although a fair amount of territory had been given up by the South in years past, it looked to many in the Confederate government as though a defensive war (such as had been fought since the beginning) could be waged indefinitely.[96]

Since Singer and Braman were called back to Texas and Fretwell and Dunn were then in command of Singer operations in Virginia and regions west of the Mississippi River, James McClintock and his old partner Baxter Watson were placed in charge of the group's Mobile facility.[97] By the spring of 1864, a reconfiguration of the original Singer torpedo design had been implemented:

> Without being very complicated, Mr. Singer's torpedo did not lead to the anticipated results after it had remained for six or seven months in fresh water, or had been lying for one month only in salt water. In the first case, the spring seems to have lost its force, and did not drive the bolt with sufficient power to cause the explosion of the caps. In the latter case, the marine worms had at-

tached themselves to the torpedo and the end of the bolt, deadening the force of the blow on the caps, and thus preventing the explosion of the charge.

This was especially observed on Admiral Farragut's attack on Fort Powell, in February 1864. A number of these torpedoes had been placed in Grant's Pass, and even west of it; the mortar-schooners of the Federal fleet very often crossed the zone of the torpedoes without causing a single explosion. On examination (of one of the torpedoes), it was found that the marine worms had formed perfect clusters on the top of the torpedo between the caps and the bolt; the pin had been withdrawn from the bolt by, as it appeared, a vessel having come in contact with the trigger-rod; the spiral spring had played, but the force of the blow had been deadened by the shells, and explosion was thereby rendered almost impossible.... Mr. Singer's torpedoes of later construction contained an air chamber. The charge consisted of from 50 pounds to 100 pounds of cannon-powder.[98]

Lt. Cdr. R. B. Bradford, an officer in the United States Navy assigned to the US Torpedo Station at Newport, Rhode Island, explained how the new configuration was employed:

> The case was made of tin, with a capacity for 50 to 100 pounds of powder; it contained an air chamber to give the necessary amount of buoyancy; the explosion was brought about by means of a plunger, actuated by a strong spiral spring, striking a rod on the inside of the case, on which was placed a percussion cap; the plunger being on the outside, the blow on the cap rod was transmitted through the medium of the case; the plunger was held with its spring in a state of tension, by means of a pin.... A cast iron weight, with a line attached to the releasing pin of the plunger, was placed on top of the case, and it was intended that this should be knocked off when the torpedo collided with a passing vessel.[99]

In a report on torpedoes commissioned by the US Corps of Engineers in 1866, the torpedo Bradford discusses was attributed to "Dr. Fretwell of Texas" and was labeled in the report as "Fretwell's Percussion Torpedo." The detailed text that accompanied a diagram of the torpedo explains why the device bore Fretwell's name: "While it seems almost certain of explosion when struck by a vessel, and has probably been the most successful one in use, it is still defective on account of the facility with which it may be exploded ... by very strong water in time of freshets. The name of the inventor is Singer, that which it bears being due to its being principally in the hands of Dr. Fretwell of Texas, who was personally engaged in planting most of them."[100]

General Rains apparently shared the Corps of Engineer's postwar opinions of the new Singer-Fretwell torpedo configuration. He stated in an unpublished manuscript on torpedo warfare, "The complexity of the outer works are objectionable, a fish can set them off, and strong currents can lean them down so as to drop the iron plate, when so arranged and thereby prematurely discharge them.... After many of these torpedoes were set in Mobile Bay they exploded daily from the tide leaning them."[101] In spite of the new torpedo's apparent drawbacks in both salt water and strong currents, production of the device commenced at Mobile and Richmond during the spring and summer of 1864, and by the end of July, over one hundred Singer torpedoes of this configuration had been anchored in Mobile Bay under the direction of operatives James McClintock and Baxter Watson.[102]

After the capture of New Orleans and the closure of the Mississippi River, only the ports of Galveston and Mobile remained open to blockade-runner traffic in the Gulf of Mexico. And since goods flowing into Galveston could be of service only to General Smith's isolated army of the Trans-Mississippi, Admiral Farragut and the officers in his command decided that the capture of Mobile would take precedence over continued operations along the Texas coast. Farragut had captured New Orleans in the spring of 1862 by forcing his way past the protective forts south of the city. He was convinced that a similar plan of attack could be used against Mobile's outer defenses. The only problem with this plan was that since 1862, the Confederates had developed reliable underwater contact mines that had proven successful on several fronts, and the admiral knew from informants and deserters that Mobile Bay was full of them. However, Admiral Farragut considered the risk of crossing Mobile Bay's submerged mine field to have been negligible:

> It became known that heavy storms, high seas, and the strong currents had torn a number of these mines from their anchorage and had driven them either on shore or out to sea. Besides this, it had leaked out that most of the mines had the Singer fuses; that these became overgrown with shell-fish and were, therefore, useless after a certain length of time. The Federals knew that these mines had been planted a long time before; therefore Admiral Farragut could reasonably expect that storm, sea and current had done their share in tearing away some of them, and that the fuses of most of those remaining were by this time unserviceable.[103]

Although Farragut perceived most of the torpedoes then anchored at the mouth of Mobile Bay to have been harmless due to their extended exposure to the elements, he nonetheless considered this mode of warfare to be "unworthy of a civilized nation" and gave orders that no quarter was to be given to anyone captured with torpedoes in their possession.[104] It was up to operatives McClintock and Watson (and the Singer personnel then assigned to the Mobile facility) to face down Admiral Farragut's armada, then seen steadily growing on the horizon off Fort Morgan.

Some two years earlier at New Orleans both McClintock and Watson (and the late Horace Hunley) had been forced to scuttle their submarine boat *Pioneer* because of Farragut's approach, and one can only assume that the two had a score to settle with this brazen Yankee admiral. With an attack on Mobile's outer defenses considered imminent, McClintock and Watson, during the final days of July, gathered up several recently fabricated torpedoes (as was undoubtedly done every few weeks or so), boarded a steamer bound for Fort Morgan, and under cover of darkness anchored them in the channel.[105] While sentries at the nearby fort walked the ramparts, McClintock and his associates perhaps took a moment and pointed to the area where their submarine boat *American Diver* was known to rest and openly grumbled that the Confederate Navy Department had never mounted a salvage operation to recover the vessel.[106] After they anchored their latest batch of torpedoes, the men returned to their Mobile facility to continue production of torpedoes slated for deployment during the weeks to come.

At about 5:30 a.m. on the morning of August 5, 1864, Admiral Farragut's attacking squadron weighed anchor and slowly steamed toward the mouth of Mobile Bay with the incoming tide. Spearheading his attacking column was the recently arrived single-turreted monitors *Tecumseh* and *Manhattan*, followed by the double-turreted USS *Winnebago* and *Chickasaw*. Behind these formidable ironclads Farragut had lashed his remaining fourteen wooden vessels together in pairs to protect his weaker ships from the furious cannon fire expected to be unleashed from Fort Morgan when his ships came within range.[107]

Lying in wait for Admiral Farragut's attacking squadron beyond Fort Morgan was the CSS *Tennessee*, the most formidable Confederate ironclad then afloat, and three smaller gunboats. At about seven o'clock the lead ships of the Federal fleet came within range of Morgan's guns, and within fifteen minutes after the first thundering shots rang out, a thick cloud of smoke had settled on the bay, which forced Admiral Farragut himself to climb the rigging of his flagship to see the progress of the fleet. In the pilot house of the CSS *Tennessee* stood Adm. Franklin Buchanan and his officers. From their vantage point they could clearly see the USS *Tecumseh* steaming ever nearer, and it was Buchanan's full intention to engage the enemy ironclad once it had passed into the bay.[108]

With a keen eye focused on the ever-nearing monitor, Buchanan gave orders that not a shot was to be fired "until the vessels were in actual contact."[109] As the anxious crew of the *Tennessee* watched the progress of the *Tecumseh* through open gun ports, they had no idea that they were about to witness one of the most extraordinary events ever seen in maritime history. Perhaps fearing that his vessel might run aground in the narrow channel, the *Tecumseh*'s captain, Tunis Craven, disregarded Admiral Farragut's orders and turned his vessel's bow westward into a submerged cluster of Singer mines.

A Confederate naval officer wrote after the war that "in a moment, and when the ships were less than a hundred yards apart [the *Tecumseh* and *Tennessee*], a muffled explosion was heard, a column of water like a fountain springing from the sea shot up beside the federal monitor; she lurched violently, her head settled, her stern went up into the air so that her revolving screw could be plainly seen, and then the waves closed over her."[110] Confederate general Dabney Maury made the following account (he seems to have been misinformed as to the actual number of casualties aboard the ironclad):

> She was leading Farragut's fleet into Mobile Bay, and running close to Fort Morgan, when a torpedo struck her. She instantly careened and went down, carrying in her one hundred and fifty officers and men. With them lies their noble Captain Craven, one of the bravest and best of American captains. As his ship was struck, Craven was by the foot of the ladder leading up to the open deck, from which he could escape. The pilot came running to get out that way; Craven stepped back, saying "After you pilot," and went down with his ship. The pilot lived to record this act.[111]

From his position in the rigging high above the deck of his flagship, the USS *Hartford*, Admiral Farragut viewed the disaster with disbelief, for within moments after his lead monitor had vanished into the depths, other attacking vessels began to show signs of hesitation. An officer who had been on the *Hartford* with Farragut gave this account: "Shot after shot came through the side, mowing down the men, deluging the decks with blood, and scattering mangled fragments of humanity so thickly that it was difficult to stand on the deck, so slippery was it. . . . The bodies of the dead were placed in a long row on the port side, while the wounded were being sent below until the surgeons' quarters could hold no more. A solid shot coming through the bow struck a gunner on the neck, completely severing head from body."[112]

It was during this moment of high drama that Farragut is reported to have shouted his immortal words, "Damn the torpedoes, four bells captain go ahead."[113] General Maury describes what the gunners at Fort Morgan witnessed: "When brave old Farragut saw the Tecumseh sink, he took the head of his fleet, hove to under fire from our guns, and lowered his boats to save those struggling men. Seeing this, noble old Dick Page, commanding the Confederate forts, ordered: 'Pass the order to fire no shot at those boats saving drowning men.' These are the chivalries which make war glorious."[114]

As the wooden vessels of the Federal fleet pushed ever forward through the submerged torpedoes, terrified sailors below decks could hear the steel detonating rods on numerous Singer mines snap against their primers as their ship's hull brushed past.[115] To this day we do not know why any of the other torpedoes did not detonate, but Farragut might have been at least partially right when he assumed that many of the mines at the entrance to the bay were defective due to prolonged submersion.[116] After the fleet passed the guns of Fort Morgan, Admiral Buchanan's flagship, the CSS *Tennessee*, and three accompanying gunboats (the CSS *Gains*, *Selma*, and *Morgan*) steamed forward and met the enemy. With odds of seventeen to four (three of the Federal vessels were turreted ironclads), soon the tiny Confederate squadron was shot to pieces and forced to surrender. By 11:00 a.m. Mobile Bay was in Federal hands.[117]

With no naval force to speak of and Admiral Farragut's squadron then inside the bay, the following orders were immediately sent to Captain Singer's torpedo facility: "Engineer Office, Mobile, August 5, 1864. Mr. McClintock, in charge of torpedoes, will proceed tonight with 30 torpedoes to or near the mouth of the Dog River. Mr. Davis and Engineer Stakes will furnish the necessary transportation. The torpedoes must be placed tonight at the locations best suited for the obstruction of the river. By order of Col. Scheliha."[118]

Although Mobile Bay was then in Federal hands and effectively sealed off as a viable blockade-running port, the city itself remained defiant until the end, for it was heavily defended by several earthen forts and gun batteries (Fort Morgan at the mouth of the bay was effectively isolated after the attack and surrendered to Farragut's forces several days later). After the loss of the bay, operatives quickly mined surrounding rivers that fed into it to bottle up the unwelcome invaders and deny them access to waterways that led to the interior of Alabama. On the day after the USS *Tecumseh* was sent to the bottom and James McClintock and his operatives had been ordered to mine the mouth of Dog River, Col. Von Scheliha sent the following message to Richmond: "Engineer Office, Mobile, August 6, 1864. To Major-General J. F. Gilmer, Chief of Engineer Bureau, Richmond Va. Sir: Fort Powell was evacuated and blown up last night. One of Singer and Company's torpedoes sunk the monitor Tecumseh. Rest of fleet passed through channel, which I have been ordered to keep open for blockade runners. The Tennessee fought the whole fleet for two hours after they passed the forts.... Am endeavoring to obstruct Dog River. V. Scheliha, Lieutenant-Colonel and Chief Engineer."[119]

Without question the sinking of the USS *Tecumseh* was the most decisive Confederate torpedo victory of the war and perhaps the most complete victory against an enemy vessel ever witnessed until that time.[120] The total destruction of such a powerful ironclad and most of its crew (within some thirty seconds) by means of a submerged mine sent shock waves through the Union navy and undoubtedly reinforced the harsh opinions already felt toward rebels then engaged in the manufacture and deployment of the mines.

With the surrender of Fort Morgan and the approaching waterways that led to the city then heavily mined, the garrison troops and citizens of Mobile slowly got used to seeing Federal warships in the bay and went about their business. Although Farragut's fleet controlled most of the waters off Mobile, he could not take the city itself until he silenced the guns at Fort Hugar and Spanish Fort, which lay directly across the shallow bay. Since there were no land forces to attack these defenses, day-to-day life in wartime Mobile could go on much as before. Farragut's fleet was then in the calm waters of the sheltered bay, so the vessels attached to the West Gulf Blockading Squadron went about their duties with little fear of attack since they had captured or destroyed the majority of the Mobile squadron on the morning of August 5.

Having captured the bay and accepted surrender of most of the vessels in the Mobile squadron, the Federal fleet came into possession of the mightiest Confederate ironclad then afloat, the CSS *Tennessee*. Housing six huge cannons and sheathed in iron armor averaging over five inches thick, this menacing steam vessel was still a formidable engine of destruction, and to let it rust at anchor on Mobile Bay was not in Farragut's plans. Although it could have been a good slap in the face to the defenders of Mobile to have turned the captured ironclad into a floating battery anchored off the city, Farragut instead decided that the vessel could best be of service with the Federal squadron then operating on the Mississippi River. His officers agreed, and Farragut sent the following orders to Lt. Cdr. E. P. Lee, who had been given temporary command of the vessel shortly after capture: "Flagship Hartford, Mobile Bay, August 28, 1864. Sir: As soon as the ram Tennessee is prepared for the voyage, you will proceed with her to New Orleans, towed by the USS Bienville. On arrival, report to Commodore J. S. Palmer, who will give you further instructions. Very Respectfully, D. G. Farragut, Rear-Admiral."[121]

As the *Tennessee* slowly made its way to the occupied city of New Orleans to be refurbished and manned by Federal sailors, Farragut could never have imagined that within months after its arrival on the Mississippi River, a Singer operative in Richmond would hatch a bold plan to destroy the ironclad, a plan that would incorporate the use of an eight-man underwater craft similar in construction to the one that had been used to destroy the USS *Housatonic* several months earlier.

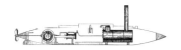

5 Operations throughout the Confederacy

IN EARLY 1863, ONE OF THE FOUNDERS of the Confederate Secret Service, Bernard Janis Sage, submitted a proposal to the Confederate Congress that outlined in part the establishment of "Bands of Destructionists" throughout the South: "Such bands would work . . . every variety of torpedo, and modes of working them defensively and offensively, submarine boats, and other contrivances of the kind. They could dash on vessels and steamboats in small boats, and capture them, or they might drench them with spirits of turpentine and other incendiary matter and burn them. . . . Blow up railroad trains to prevent reinforcements and supplies during a battle. . . . Such bands and the kind of warfare contemplated, must be authorized, so that persons engaged may have military status, and be protected by the government."[1] All these forms of warfare had been carried out at one time or another by various members of Captain Singer's Secret Service Corps. Certain members in both the McDaniel and Courtenay Secret Service Corps were engaged in this form of sabotage along the Mississippi, so it is not surprising that some members of Captain Singer's group may well have been assigned the task of burning the enemy's "vessels and steamboats."[2]

The following report filed in early October 1863 gives us some idea about when the intentional burning of enemy steamboats commenced along the Mississippi River:

> Office of Chief Quarter Master, Saint Louis, October 5, 1863. Major-General H. W. Halleck, General in Chief, Washington, D.C. General: The continued destruction of steamboats, by fire, on these waters is assuming a very alarming feature. Unquestionably there is an organized band of incendiaries, members of which are stationed at every landing. It is a current report here that the Confederate Government has secretly offered a large reward for the destruction of our steamers. Already some fourteen first-class boats have been burned, and this is equivalent to ten percent of the whole river transportation. Increase of watchmen and extra vigilance do not seem to arrest this insidious enemy. The incendiary, when it serves his purpose, becomes one of the crew, and thus secures himself from detection. I apprehend that there are disloyal men in disguise in the employ of every steamer, and it will be difficult to examine them. Very respectfully, Robert Allen, Chief Quartermaster.[3]

Joseph W. Tucker, a Saint Louis native, was the man in charge of this covert boat-burning operation and a member of the Confederate Secret Service whose affiliation with that organization can be traced to the early months of 1862.[4] Operating from headquarters established at Mobile, Tucker and several associates recruited agents and coordinated boat-burning operations from Saint Louis to New Orleans. The following letter from Tucker describes the mission of this loosely organized group of saboteurs (that apparently included members from all three of the known secret service corps, McDaniel's, Courtenay's and Singer's):

> Spotswood Hotel. March 14, 1864. Confidential Statements for the President alone. A deputation, under the authority of the order, was sent to confer with me at Mobile in relation to the destruction of the enemy's marine service, together with armories, arsenals, depots of stores, etc. etc., as a means of weakening and paralyzing the military strength of the Federal Government.... Our future plan, if sanctioned and aided by the government, embraces the destruction of that transport service upon which Grant must rely in the great coming struggle of the spring campaign; a week ago we burnt $500,000 worth of hay at the Memphis wharf, to embarrass Sherman; not long since Colt's pistol and gun factory became an earnest of what can be done. We design to strike a blow on the same day, at many points, that will paralyze the foe.[5]

On March 23, 1864, Jefferson Davis approved a payment of forty-six thousand dollars in gold to Tucker for continued covert operations along the Mississippi River. With a substantial amount of gold then at his command (under the watchful eye of Gen. Bishop Polk),[6] sabotage of boats on the river and burning enemy supply depots behind their lines continued throughout the summer of 1864.[7]

Although Joseph Tucker seems to have been in overall command of operations along the Mississippi and its tributaries, Gen. Bishop Polk occasionally sent various operatives on special missions into occupied Kentucky and Missouri. General Polk himself selected Henry Dillingham to undertake just such a special mission.[8] Dillingham had been a member of the second *Hunley* crew during early October 1863 and had been ordered with Lt. George Dixon to "Mobile and return on business connected with the submarine torpedo boat" shortly after its accidental sinking in the Cooper River.[9] Henry Dillingham served as a crewman and perhaps engineer aboard the *Hunley* during the fall of 1863.[10] He was likely a member of the Singer organization, because General Polk during the spring of 1864 personally selected Dillingham and Bill Noland, a Confederate Secret Service agent, to venture deep into Union-controlled Kentucky to burn enemy supply depots and river transports.[11]

During the summer that followed several troop transports and Federal supply depots throughout the Mississippi region fell prey to the "Boat Burners," a name coined by Federal authorities and detectives assigned the task of ferreting out these rebel agents.[12] Capt. Thomas Courtenay wrote a letter in the fall of 1864, noting how popular his coal torpedo had become with Confederate saboteurs operating behind Union lines: "By a very simple invention I have been enabled to do much injury to federal steam property and have proved beyond a doubt that iron-steel built steam vessels can be more easily destroyed than by cannon or conical shell.... They have destroyed many steamers on the Mississippi River and a few months ago blew up the new gunboat Chenango at Brooklyn, New York."[13] (At the time of the *Chenango*'s destruction, with great loss of life, military authorities thought that the boilers had been faulty, a false assumption that caused the naval boiler inspector, Second Assistant Engineer S. Wilkins Cragg, to be dismissed for "neglect of duty.")[14]

In early August 1864, while Singer operative James McClintock was mining the entrance to Dog River near Mobile, Henry Dillingham, Edward Frazier, and Thomas Clarke arrived in Richmond and took rooms at the Spotswood Hotel, where John Fretwell was then on duty. We do not know why the three men had recrossed enemy lines at Memphis and journeyed to Richmond, but Henry Dillingham was apparently seeking a bounty from the government for burning a medical supply depot in Louisville, Kentucky.

A captured Confederate agent, who himself had been engaged in boat burning along the Mississippi River, testified that Edward Frazier, like Henry Dillingham, had been employed as a steamboat engineer for some years prior to the South's secession.[15] The agent also revealed that the bounty on Federal vessels destroyed was to be 50 percent of the ship's value. To fix a price and make settlements with the saboteurs, the government agreed that a destroyed vessel's worth would be determined by its value according to northern newspapers. It was therefore the responsibility of the agent who sought a bounty to gather related news clippings that discussed his act of sabotage and turn them over to the proper authorities for payment.[16]

Within days after Dillingham, Frazier, and Clarke had arrived in Richmond, they were granted a meeting with the Confederate secretary of war, James Seddon. Edward Frazier testified some months later concerning how the three agents were received at the Confederate capital:

> I was in Richmond ... until about the 25th or 26th day of August 1864. I there had an interview with the Secretary of War, the Secretary of State, and Mr. Jefferson Davis. Thomas L. Clarke, Dillingham, and myself went there in connection with the boat burning, and put in claims to Mr. James A. Seddon. Mr. Clarke introduced me to him, and he said that he had thrown up that business, and it was in the hands now of Mr. Benjamin [the Confederate secretary of state].

We went to see Mr. Benjamin. Mr. Benjamin looked at the papers, asked me if I was from St. Louis. I told him I was. He asked me if I knew anything about these [the claims]. I told him I did, that I believed they were right.... He said for me to call back the next day with Mr. Clarke and Mr. Dillingham. I called back next day and he said he had shown those [the claims] to Jefferson Davis, and he wanted to know if we would not take $30,000 and sign receipts in full. We told him we would not do it. Then he said if Mr. Dillingham was to claim this in Louisville [the burning of a medical supply depot some months earlier with the assistance of Confederate agent Bill Noland], he wanted a statement of that for Louisville.

We went back to the hotel and drew up a statement of it. I wrote the statement out myself. It read that Mr. Dillingham had been hired by General Polk, and sent to Louisville expressly to do that work [burn federal supply depots]. I signed Mr. Dillingham's name to it. That was given to Mr. Clarke. Mr. Clarke took it over to Mr. Benjamin and we made a settlement with him. We made the settlement for $50,000, $35,000 down in gold and $15,000 on deposit.[17]

Fortunately, an accounting of this payment in gold has recently been confirmed from an entry in a Confederate Secret Service account book presently in the collection of the Chicago Historical Society. An entry dated August 20, 1864 (the date Henry Dillingham and his associates filed their claim for the destruction of the medical supply depot at Louisville), confirms that Jefferson Davis, on that date, approved a payment of thirty-five thousand dollars in gold for "Secret Service."[18] Known as the "Chicago Remnant" to students of the Confederate Secret Service, this ledger book kept at the Confederate State Department at Richmond recorded payments to individuals engaged in covert operations throughout North America and Europe. From entries in the secret ledger book, it was obviously picked up in Richmond as a souvenir during the closing days of the war; by 1874 the book had found its way into the possession of an Illinois man who had purchased a drugstore.[19] Unfortunately, the remaining pages (as well as several that had already been written on) were used for many years to record various inventory at the drugstore, and all that remains of the "remnant" today are several confusing pages that outline payments for secret service activities yet to be discovered.

Edward Frazier also testified regarding a secret mission for the newly arrived trio that had apparently been hatched by Jefferson Davis himself.

While there Mr. Benjamin said Mr. Davis wanted to talk to me. I went in, Mr. Benjamin and myself and Mr. Davis sat there and talked. The conversation went on a bridge between Nashville and Chattanooga—the Long Bridge they called it. Mr. Benjamin, I believe was the one who mentioned it first. Mr. Davis wanted to know if I knew where it was. I told him I did; but I do not; I have never been there. He wanted to know what I thought about destroying that bridge. He said that he had been thinking about it, they had been thinking about sending someone to have it done.

I told him I did not know what to think about it. He said I had better study it over. I finally told him that I thought it could be done. Mr. Benjamin first made the remark that he would give $400,000 if that bridge was destroyed and wanted to know if I would take charge of it. I told him that I would not have anything to do with it unless all passes were taken away from the men down there, that nobody should be allowed to go up anymore, and they said it should be done. Then the conversation turned to the burning of steamboats....

I asked Mr. Davis moreover, if it made any difference where the work was done. He said that it did not; it might be done in Illinois or any place, such as railroad bridges, commissary and quartermaster's stores, anything appertaining to the army, but as near Sherman's base as possible, that Sherman was the man who was doing more harm than anybody else at that time.... I saw an order issued in the paper next day taking away all passes issued on or before the 23rd of August.[20]

The three traveling companions wasted little time in putting their plans into motion, for on the same day that Frazier had accepted the bridge-burning assignment from President Davis, the following request for passes was submitted to the secretary of war:

Richmond, Va., August 22, 1864. The Honorable James A. Seddon, Secretary of War. Dear Sir: The following named persons are members of an Association the object of which is the destruction of federal property within their lines. Edward Frazier, Henry Dillingham, Joseph C. Dillon, Thomas B. Smith, Harrison Fox, Owen Jones, Hardy Wilson, John Rawson, William R. Erwin, Clay Thomson, John G. Parks, George Burton and Thomas L. Clarke. Thomas L. Clarke is our agent and attorney for the transaction of our business with you, as it is our wish that you should take the control of this matter in your own hands. Our reason for this is that we wish to make our communications to you direct. By granting the necessary papers to the parties named it will show them that they are supported by you. All the parties named are actually employed in this business which I will vouch for. Respectfully yours, Edward Frazier.[21]

By the time Jefferson Davis met with Edward Frazier to discuss the burning of the Long Bridge near Nashville, Gen. William T. Sherman was laying siege to Atlanta. As the siege began, Sherman's main concern centered on maintaining his supply trains, for it was well known that if these daily trains carrying munitions and rations through occupied Tennessee were ever cut off, he and his army would be forced to retreat northward. Jefferson Davis wanted to cut off these much-needed supplies at their source, apparent from his interview with Frazier.

Both the Courtenay and McDaniel Secret Service Corps had detailed several agents along the Mississippi River in prior months to conduct sabotage operations against Federal steamships and supply depots.[22] Several of Singer's operatives may also have been assigned this task, since Henry Dillingham may well have been inducted into Captain Singer's unit. Unfortunately, a complete roster of Singer operatives may never be compiled. An extremely rare McDaniel's Secret Service Corps document dating from late August 1864 notes posts assigned but omits names of agents: Richmond, 12; Kentucky, 4; Mobile, Alabama, 3; Augusta, Georgia, 2; Mississippi, 2; Lynchburg, Virginia, 2; Missouri, 1; Tennessee, 1; Pittsylvania County, Virginia, 1.[23] Personnel were dispersed in both Confederate and Federal territory (Kentucky, Tennessee, and Missouri were all in Federal hands), and Singer operatives were most likely dispersed in a similar manner. The following are known locations of Singer personnel for the same period: Richmond, 2 (Fretwell and at least one assistant); Wilmington, North Carolina, 1 (James Jones); Mobile, 4 (McClintock, Watson, King, and Andrews); Houston, 4 (Singer, Braman, Bradbury, and Longnecker); Shreveport, 1 (Dunn). The whereabouts of Frary, Hirshberger, and Tucker are not known for the late summer of 1864. Captain Singer's group had about twenty-five members,[24] so Edward Frazier's list could have included several Singer agents.

Captain Courtenay's recently perfected coal torpedo seems to have been issued to numerous Confederate agents engaged in sabotage behind enemy lines and along the Mississippi River.[25] Even southern political agents operating in Canada (who did not engage in acts of sabotage) were issued coal torpedoes, so Henry Dillingham and his two associates may also have been given such devices since they were to destroy enemy steamships along the Mississippi River.[26]

With passes issued (thirteen in all) to Frazier, Dillingham, and Clarke by the secretary of war and a voucher for thirty-five thousand dollars in gold (signed by the secretary of state to be drawn from the treasury at Columbia, South Carolina), the three made plans to leave Richmond as soon as possible. They boarded a southbound train on the first leg of their journey to Columbia, South Carolina. All we know about the group's brief detour to the Confederate Treasury at Columbia is that Henry Dillingham and his partners, clad in long dusters, did receive his bounty.[27] We can only imagine how the chief of the Confederate Treasury reacted as he watched numerous gold ingots entrusted to his safekeeping disappear into a dusty old carpetbag spread open on his desk. And why such a large amount of gold was given to the strangers was probably never revealed in the documents Henry Dillingham produced.

The three men continued to Mobile, where they spent nearly two weeks planning the next leg of their hazardous journey while assembling additional personnel for their mission into occupied Tennessee.[28] William Murphy, a Mobile boat carpenter (who had been recruited for similar missions in the past) testified about how he got involved with the three saboteurs: "I worked until about four or five days after Fort Morgan fell, when a man named Frazier came. I was introduced to him, I knew a man named Henry Dillingham, who was with Frazier, and from him I found out that Frazier was an agent—him and a man named Clarke—for the Confederate Government, and had men employed to burn steamboats. He [Dillingham] had been on to Richmond, I believe, to collect some money for work he had done. I got recommended by Henry Dillingham and one or two others. He engaged me to come out with him and help to burn up something."[29]

While awaiting various members of the group to join them at the Battle House Hotel in Mobile (including John Rawson, Bill Erwin, John Parks, and Mitchell Pete, whose name does not appear on the August 22, 1864, list of saboteurs submitted to the secretary of war),[30] Henry Dillingham undoubtedly interacted with both James McClintock and former fellow *Hunley* crew member Lt. William Alexander. For some six months William Alexander had presumably been working on his design for a breech-loading cannon at the Park and Lyons machine shop, and Dillingham, being an engineer himself, would have been curious to see what progress Alexander had made since being called back from Charleston.

With the core group assembled by mid-September, the Tennessee-bound saboteurs boarded a train and headed for Mississippi. William Murphy testified that "in Mobile, before we started, the gold was in three satchels and a pair of saddlebags. . . . We had passes that we showed at all the military posts signed by Secretary Seddon." From Mobile we went "to Grenada, from there to Panola, Mississippi, and from there to Senatobia, from there to Fletcher's, near Memphis, in a wagon, the wagon was chartered and paid for by Frazier—it was a two horse wagon."[31]

Fletcher's refers to "Fletcher's house [a Confederate safe house run by an agent named Fletcher that was to become group headquarters] about 18 miles from Memphis, on the Hernando Road."[32] Murphy's testimony continues:

> When we got to Memphis the passes were all given to Clarke. . . . We arrived at Fletcher's about nine, and we got up about six next morning, and started for Memphis about eight. Rawson and I together over the railroad track (we walked in); Thorpe and Pete walked in on the Hornando Road; Frazier came in a wagon; Dillingham and Erwin stayed there. . . . We met at a little clothing store kept by Pat Connally. . . . The next day we came up on a boat to St.

Louis—Frazier came too at the same time. . . . The others stayed in Memphis and were to come up as soon as they could. . . . Every man was to go off wherever he pleased. . . . I think Rawson is in Pennsylvania.[33]

Murphy apparently had never been informed about the mission to burn Long Bridge, as there is no mention of it in the original handwritten text. Henry Dillingham may have been given the task of destroying the bridge, for Murphy stated that Dillingham and Bill Erwin stayed at Fletcher's safe house near Memphis after the others had gone. Murphy may have had no idea what Dillingham's mission was or where he went after the others had entered occupied Memphis. Nothing more is known of Henry Dillingham's whereabouts for the rest of the conflict. Some weeks earlier news that Sherman had taken Atlanta reached Mobile, which may have caused the group who were to destroy the bridge to change their plans.

After the war Adm. David Porter wrote about Confederate sabotage:

> By assiduous watching of small boats crossing the Mississippi, I succeeded in capturing a package of dispatches which fully explained the organization of a corps of Confederate torpedo-setters, together with the names of the parties concerned [all members of Captain Singer's unit], and their commissions signed by Mr. Mallory, "Secretary of the Navy." This old sea-dog, not having any vessels wherewith to operate on the ocean, except the one commanded by Semmes [the CSS *Alabama*] organized a body of horse-marines to patrol the shore, who were directed to sink, burn, and destroy every Union vessel on the Mississippi and its tributaries by means of the new style of infernal machines.
>
> In order to circumvent these machinations, I appointed a corps of detectives to travel in all merchant-steamers, and win the confidence of the rebel operators. Some of the latter ended their career very suddenly. The general order which I thought necessary to issue at that time recites the reckless character of the people with whom we had to deal. . . . One very valuable vessel was destroyed by these infernal machines. She was used as a "wharf-boat," or a store-ship, at the Mound City Navy-Yard, was six hundred feet long and sixty feet wide, and filled with stores for the fleet. Notwithstanding the greatest vigilance was exercised, a torpedo resembling a lump of coal was introduced on board, and the vessel was destroyed by the fire which took place after the explosion.
>
> At the time of this occurrence, my flag-ship, the Black Hawk, was made fast to the wharf-boat, and the first notice I had of her danger was a slight explosion, then the whole vessel was immediately wrapped in flames. Here is a torpedo beneath the notice of the Bushnells and Fultons, yet sufficiently effective in its particular line. It would perhaps require a subtle casuist to determine how far such contrivances are justifiable in war.[34]

While Henry Dillingham and his associates were in Mobile Edgar Singer wrote to the assistant adjutant general:

> Houston Texas, August 14, 1864. Major Barton, Assistant Adjutant General. Sir: I have the honor to report that myself and three of my company are on duty at this point constructing by order of General E. Kirby Smith a torpedo boat. . . . The following is her dimensions length 114 feet breadth of beam 14 ft . . . double fluid boilers 24 ft. long with two direct acting oscillating engines 12 x 16 in. [diameter of the pistons], diameter of propeller 6 ft with 10 and ½ ft. pitch to make 120 turns per minute. . . . 2 more of my company are on duty in Trans Miss. Department the rest are at Mobile and vicinity under direction of James McClintock constructor of torpedoes and Greenleaf Andrews as chief operator during my absence. . . . General Smith has ordered some of my torpedoes planted in the Mississippi River which I am looking to hear from daily. Your Obedient Servant E. C. Singer.[35]

While Singer's report was making its way to the capital, Henry Leovy was preparing for meetings in Richmond with the secretary of war. Some months earlier, Leovy was offered, with the rank of major, the position of Confederate states commissioner for southwestern Virginia.[36] Leovy's biography states that "Jefferson Davis appointed him as commissioner to examine and settle disputes between civil and army authorities in southwestern Virginia concerning the large number of arrests of suspected spies and deserters."[37] During these investigations Major Leovy apparently uncovered a network of traitors and informants code-named "The Heroes of America." Leovy had convinced the secretary of war that further investigations were needed, for James Seddon immediately placed two undercover operatives, John Williams and Thomas McGill, under Leovy's authority.[38] With his detectives in tow, Major Leovy returned to southwestern Virginia and briefed the men on the various code words, grips, and signs used by members of the treasonous group that he was determined to expose (these various code words, etc., had been revealed to Henry Leovy during the interrogation of a "suspected party under military arrest on other grounds" who wished to make a deal in return for his freedom[39]). His undercover investigation evidently paid off, for within three weeks he was filing a report from his headquarters in Pulaski County, Virginia.

During this time Singer operative Baxter Watson, then on duty in Mobile, was preparing to contact government officials regarding construction of another submarine boat. McClintock, Hunley, and Watson had experimented with electricity while fabricating the *American Diver* two years earlier. According to McClintock, "There was much time and money lost in efforts to build an electro-magnetic engine for propelling the boat."[40] Electric motors (known during the Victorian Age as "electro-magnetic engines") were commercially produced prior to and during the American Civil War, and information on how to build such devices can be found in copies of *Scientific American* dating as far back as the late 1840s.[41] Watson, McClintock, and Hunley had built such a motor to propel the *American Diver* in late 1862, but they "were unable to get sufficient power to be useful."[42]

The Singer group decided that one of them would journey north and simply purchase an electro-magnetic engine from a northern supplier. Although Watson and those committed to the project seem to have thought out their plans, they could not secure sufficient funding from within the group, prompting Watson to submit the following plea to Jefferson Davis:

> Mobile, October 10, 1864. Hon. President Davis. Sir: Being the inventor of the Submarine boat that destroyed the Yankee vessel "Housatonic" in Charleston Harbor in February last and being unable to build another. I have concluded to lay the matter before you and ask your assistance and influence in the matter as I have exhausted all the capital that I had in building and experimenting with that one which was lost in Charleston. . . .
>
> That boat was a complete success as far as the boat was concerned, but was a complete failure from mismanagement by those who had her under control. A boat of that description I am satisfied after three years' experience can not be used successfully without Electro Magnetism as a motive power, the air in the boat will not sustain so many men long enough for the time required, or if it did, the labor necessary for the successful operation of the boat is more than they can possibly endure.
>
> Knowing this: I have tried to procure an Electro magnetic engine, but so far have not accomplished it. I can procure an engine of this description by going to New York or Washington City, but the amount of five thousand dollars in exchange, necessary to defray the expenses in Confederate money is more than I can raise, as moneyed men are afraid to invest lest the war is over before they could realize their profit and consequently they would lose. I think I can raise the money among my fellow mechanics if the government will give them the privilege of buying cotton and sending it out to a foreign port, by that means the money here will be almost equivalent to exchange and it is for that purpose that I have addressed this communication to you knowing the delays that would undoubtedly occur by applying to other officials.
>
> I can fully satisfy you that this is of much value to the Confederacy and would like to have the government take the matter in hand, but if not so disposed, I hope it will at least allow me the privilege and facilities for doing it. If you desire it I will be happy to give you a full description, and all the necessary proof of the utility of the enterprise and my sincerity in its completion, hoping you may see the advantages to be gained is the wish of your humble servant Baxter Watson. Care of Park and Lyons.[43]

Davis summarized the communication, endorsed it on the back, and then quickly forwarded it to James Seddon for consideration. Seddon knew virtually nothing about submarines, so he apparently sent Watson's proposal to the commander of the Engineering Department, Gen. J. F. Gilmer: "20 October, 64. To Engineer Bureau: Can you relate its performance at Charleston? J.A.S [James A. Seddon]."[44] Within days Watson's proposal was returned to the War Department with the following words scribbled on the back: "Engineer Bureau. 27 October 1864, Respectfully returned to Honorable Secretary of War.... One or two crews were lost.... Further expenditures on similar constructions are consequently not recommended. J. F. Gilmer, Major General and Chief of Bureau."[45] Watson, however, never gave up hope and continued to petition various members of the Confederate military for funding at least into early January 1865.[46]

Meanwhile, ironclad torpedo boats were being constructed in Houston. From early July through late October 1864 not a word regarding the construction of either of the two iron monsters appears in any Trans-Mississippi Department or Department of Texas, New Mexico, and Arizona order book. The silence was broken on October 27, 1864, when Gen. John G. Walker (Walker relieved Magruder as commander of the Department of Texas on August 17, 1864; Magruder was given command of the Department of Arkansas) filed the following report:

> Head Quarters District of Texas, New Mexico and Arizona. October 27, 1864. Colonel Smith Commanding Defenses of Galveston. Colonel: Mr. Close the owner of the foundry at Galveston is under contract with the government to furnish an engine for the torpedo boat now being constructed under the superintendence of Mr. Dunn. As an excuse for non compliance with his contract, Mr. Close alleges that he is engaged in other government work principally for the Engineer Department. As the torpedo boat promises good results in the defense of Galveston harbor, it is desirable that no embarrassment be thrown in the way of the speedy completion of the engines above referred to. It is thought that Mr. Close is not inclined to fulfill his contract under any circumstances and I desire that you will inquire into the matter and urge upon him the fulfillment of his agreement with Mr. Dunn, and see that no obstacles to the early completion of the work are thrown in the way. Very respectfully, your obedient servant, J. G. Walker, Major General Commanding.[47]

David Bradbury and Robert Dunn had been hard at work over the past several months, and military headquarters looked favorably on the fruits of their labors. Another similar report regarding the Houston torpedo boats was not filed until six weeks later. Thus, we return to events then taking place in southwestern Virginia.

Captain Singer had by early November 1864 returned to the Confederate capital.⁴⁸ While Singer was in Richmond, he was most likely meeting with and working closely with John Fretwell (who had been assigned to work with General Rains's Torpedo Bureau several months earlier). The only thing known about when Singer arrived and how long he remained are noted in this brief entry in a book of endorsements: "Number" 406, "Name of Writer" E. C. Singer, "Where Written" Richmond, "Date" November 11, 1864, "Contents" Report upon operations of torpedo company.⁴⁹

This summary regarding "operations of torpedo company" is all that remains of perhaps Edgar Singer's most comprehensive report filed during the war years. With various operatives stationed throughout the Confederacy on assignments ranging from the construction of huge ironclad torpedo boats, to the laying of underwater contact mines, and the assumed burning of river transports and depots behind enemy lines, the lost report undoubtedly contained information on personnel and covert operations that will never be known. The mystery is further compounded from several lines in an official Engineering Department document penned on the same day Captain Singer submitted his report. The bulk of the communication seems to center on James Jones's transfer back to Colonel Shea's artillery battalion in Port Lavaca:

> Head Quarters 3rd Corps, Army of Northern Virginia. Engineer Department, Petersburg 11th November, 1864. Brigadier General W. H. Stevens, Chief Engineer. General: I have the honor to submit for your action the following James Jones, a private in Shea's Battalion Heavy Artillery Texas Coast was detached some eighteen months ago for duty with E. C. Singer and Company in the manufacture and use of the Singer Torpedo. With others he was sent by Major General Magruder to Lieutenant General E. Kirby Smith, and by him to this side of the Mississippi River.
>
> He has been in the employ of the Engineering Department in this vicinity and Wilmington, but under the recent order I presume he must be ordered back to his Company, supposed to be in Texas. His service as a mechanic induced me to place him in charge of some repairs of Steam Engines.... Jones holds your own statement or certificate of his being under employee of the Department, but has nothing to show his official status. He has left them in the Trans-Mississippi Department for fear of capture while crossing river. He is entirely reliable and one who can be with safety entrusted with any secret undertaking.
>
> He was under the impression that he had been put by the Engineer Bureau on the same footing with Singer, Dunn and Braman, but had no written notice to that effect, though told by Mr. Dunn that he had been assigned to Captain Singer's Secret Service Company. He prefers to be sent to the Trans-Mississippi Department as his house and people are there, and I hope it will be in your

power to have him sent there. Very Respectfully. J. N. Williams, Lieutenant Colonel Engineers.⁵⁰

Judging from the numerous endorsements that appear on the back of this document, there was some confusion about where James Jones should be detailed. One of these summaries reads, "James Jones an employee of the Engineer Department. Is he a member of an Artillery Company or of a Secret Service Company from Texas?" In response the following appears: "The services of James Jones are no longer needed by Colonel Williams, I therefore recommend that orders be given him to return to his Company. J. F. Gilmer, Major General and Chief of Bureau."⁵¹

With General Gilmer's handwritten endorsement appearing on the back, the document was forwarded to both the Adjutant and Inspector General's Office and the secretary of war by November 28. James Jones had met with Captain Singer at some point during this period and received orders to return to Texas, with a bold Singer plan to recapture enemy-controlled waterways in the Trans-Mississippi theater. With secret orders presumably in hand, James Jones was ordered to Houston by the Confederate inspector general on December 8, 1864.⁵²

While Captain Singer and various officers attached to the Engineering Department were deciding where James Jones should be assigned, John Fretwell was receiving revised orders. Gen. Gabriel Rains summarized his department's use of both the Singer torpedo and subterra shells (land mines) and discussed the temporary redeployment of Fretwell to Goldsborough, North Carolina:

> Torpedo Bureau, Richmond, Va., November 18, 1864. The Honorable James A. Seddon, Secretary of War. Sir: I have the honor to state that notwithstanding the vigilance of the enemy we have managed, from time to time, to transfer to their rear torpedoes.... It has had one good effect, in causing the enemy to watch the river-banks with thousands of their soldiers, who otherwise might be employed against us. We have relied somewhat necessarily upon the "Singer Torpedoes," which were located at spots visited by the boats of the enemy.... Our operations have been mainly directed to the James, Pamunkey and Chickahominy Rivers, and some attempts made in Appomattox with torpedoes.... Our efforts for the defense of this place [Petersburg] have been directed lately to planting subterra shells between our lines of abatis at our works commanded by General Barton. We have planted at this date 1,298.... I have sent Dr. Fretwell, an expert with Singer torpedoes, to Brigadier-General Baker at Goldsborough, who may be able to effect something in Tar River at or near Washington [NC]. Very respectfully, your most obedient servant, G. J. Rains, Brigadier-General, Superintendent.⁵³

Apparently Fretwell took with him to North Carolina several dozen recently fabricated Singer-Fretwell torpedoes, a timely decision that would spell disaster to a small Federal flotilla soon to be dispatched up the narrow Roanoke River. By the time Fretwell had been ordered to Goldsborough with his torpedoes, news that Abraham Lincoln had won a second term in office was well known throughout what was left of the ever-shrinking Confederacy.

Because the Democrats had been soundly defeated, talk of peace negotiations with the South was by then a dead issue, for the Confederates knew that the Lincoln administration was determined to wage war until all signs of resistance had been crushed. In the few months that had elapsed since the summer of 1864, Atlanta had been captured, Mobile Bay had been effectively sealed off to blockade-runner traffic, and General Lee's Army of Northern Virginia was pinned down in muddy trenches outside Petersburg. Since Lincoln had been returned to office for another four-year term, the South had no other course than to fight to the end.

As both armies prepared to spend the winter months bivouacked before Petersburg in rain-soaked trenches often oozing with filth, General Grant and the officers in his command devised a bold plan to cut off Lee's southern supply trains. By the fall of 1864 most of the railroads leading to the Confederate capital had either been destroyed or were then in the hands of Grant's army. Although Grant had been successful in severely disrupting the routes by which Lee's army obtained their much-needed supplies, he had not yet been able to cut off the all-important supply route between Richmond and the blockade-running port city of Wilmington. Grant decided that the Weldon Railroad, the last lifeline linking the two rebel strongholds, had to be severed.[54]

On December 1, 1864, some two weeks after General Rains had dispatched Fretwell to North Carolina with several dozen Singer torpedoes, the commander of the North Atlantic Blockading Squadron issued an order to destroy the Weldon Railroad bridge at Rainbow Bluff on the Roanoke River. The following report describes the events that transpired after the small Union flotilla entered a stretch of river that Fretwell and his assistants had visited some days earlier:

> U.S.S. Wyalusing, Roanoke River, North Carolina, December 11, 1864. Admiral: I have the honor to make the following report of the operations of the naval part of the expedition to capture Rainbow Bluff, in obedience to your orders of the first of December. . . . I started up the river from Plymouth at 5 p.m. December 9, in the Wyalusing, leaving the Shamrock to guard Plymouth. The Otsego, and Valley City, and tugs Belle and Bezely, and the picket boat number 5 started at the same time, and I left orders for the Chicopee, Ceres, and Commodore Hull to follow as soon as they should arrive at Plymouth.

After steaming slowly up the river . . . I came to anchor for the night. . . . I made signal for the other vessels to do so also. In obedience to this order the Otsego had stopped her engines and was just about to let go of her anchor when a torpedo exploded under her on the port side, and shortly afterwards another exploded under her forward pivot gun, which was thrown over the deck by the concussion, the two explosions injuring her so badly that she sank in a few minutes, her spar deck being about three feet under water.

Fortunately no one was killed . . . and with the exception of a few slight scratches no one was injured. The Otsego had spars rigged out ahead of her to which was fastened a net for the purpose of catching the torpedoes, and two were found in the net after she sank. She must have stopped directly on top of a line of these infernal machines. Six were picked up after the explosion. . . . In the morning I determined to send the Bezely to Plymouth to find out what the army were doing and to get rations from the Shamrock for the Otsego's crew. I also wished to get up a coal schooner to take the guns from the Otsego.

I got the Bezely along side of the Wyalusing and sent Acting Assistant Paymaster Sands on board to take the dispatches to Plymouth, and then sent the tug to the Otsego to get some men and an officer. When the Bezely got within a few yards of the Otsego another torpedo exploded under her, and she went down right along side the Otsego. By this explosion 2 men were killed, but none of the officers. While at this place we fired with musketry and great guns into the woods on the banks. . . . We however, heard nothing from the army that day, and in the evening I sent picket boat Number 5 with an armed launch belonging to the army up the river to try to get some news from the army forces. . . . The boats returned this morning with intelligence, obtained from Negroes on the river, that the river is full of torpedoes, these being as many as forty in one place on the river.

These torpedoes are made on some new plan with an air chamber. . . . They are very sure, and every one we have picked up was in good condition, two of them bursting as we were hauling them ashore, but hurting no one. . . . After the destruction of the Bezely we had the river dragged all around, and six more torpedoes were picked up, all right by the Otsego. . . . I have determined to send First Assistant Engineer S. C. Midlam to you with dispatches, to inform you of our situation and ask your advice.

In the mean time we will continue up the river, though our progress will be very slow, for we shall have to drag all the way before us with small boats, that being the only way we can proceed with the vessels with any safety. I have taken out most of the guns from the Otsego, and am continuing the work. . . . When we advance any further it will be necessary that we have more vessels; otherwise we will be unable to patrol the river below us, and consequently the rebels will be able to lay more torpedoes and blow us up on our return. Very respectfully, your obedient servant, W. H. Macomb, Commander, Commanding District Sounds of North Atlantic Squadron.[55]

With two of his vessels then at the bottom of the Roanoke River, Commander Macomb moved toward his objective at Rainbow Bluff with extreme caution, for he knew that many more infernal machines lay hidden in the muddy waters before him. As news of the Federal advance up the Roanoke spread, Confederates quickly recognized the operation for what is was and immediately reinforced their positions near the Weldon Railroad bridge at Fort Branch.[56]

In 1880, Lt. C. W. Sleeman of the British Royal Navy wrote about Fretwell saving the bridge: "A notable instance on the effect of torpedoes on the war was the saving of the Weldon line of communication in December 1864. The Weldon Railway was the principal artery of communications to Richmond for the Confederates. To intercept this, by destroying the railway bridge, a fleet of nine Federal gunboats was sent up the Roanoke River; when nearly arrived at their destination, and though every precaution in the shape of bow projecting spars, creeping, etc., was taken, the vessels were either sunk or severely injured by submarine mines. Thus the expedition ended in a most disastrous failure."[57]

Apparently Lieutenant Sleeman considered John Fretwell's complex torpedo defenses of the Roanoke River to have been a nautical first worthy of history's attention, for he emphasized that "the Weldon line of communications to Richmond" had been saved from destruction only because "submarine mines" had been anchored in the path of the attacking foe. Thanks to Fretwell and his associates, Wilmington continued to be one of the principal sources of supply for Lee's army until the closing weeks of the war.[58]

About this time Robert Dunn was in the process of securing an old locomotive steam engine for one of the huge torpedo boats being constructed near Houston. Some days earlier, Dunn and an assistant named Clark had journeyed to Columbia, Texas, and examined two government-owned locomotive engines with hopes that one of the huge soot-covered monstrosities might be of service.[59] As a result, the commander of the Department of Texas, New Mexico, and Arizona wrote this order: "December 12, 1864, W. W. Morris Superintendent Texas Rail Road: I have that there is an old locomotive engine in your possession belonging to the government. As it is said to be unfit for Rail Road purposes, it is proposed to use it for the boat being constructed under the superintendence of Mr. R. W. Dunn, to whom you will please turn it over. Signed, Major General Walker."[60] Early December seems to have been a busy time for the Singer operatives engaged in building the ironclad torpedo boats, for on the same day that General Walker wrote this order, the following summary was transcribed in the book of endorsements at Houston headquarters: "R. W. Dunn, Chief Constructor Torpedo Boat, Houston, December 12, 1864. In regard to Iron wanted for construction of boat . . . Reply referred to Captain Lubbock Commanding Marine Department, who will order the iron referred to within brought to Houston without delay and turned over to Mr. Dunn."[61] Although the quantity of iron requested was never specified, Dunn and the ship carpenters assigned to his project were probably then preparing their vessel's hull to receive its iron skin. From deserter testimonies now on file at the National Archives we know that "Rail Road Iron" was to have been used to sheathe the vessel's hull,[62] so the iron ordered to be delivered to Dunn was in all likelihood various lengths of old iron railroad track.

We can assume from additional directives found in surviving Texas order books that construction on the other ironclad torpedo boat was then taking place at Houston, on Buffalo Bayou, at a location known as Lubbock's Mill.[63] The torpedo boats were being constructed at different locations to maximize the use of local facilities, as well as keep one of the boats farther inland in the event that Galveston was again captured by the Federals.[64]

From documentation appearing in the Department of Texas, New Mexico, and Arizona order books, the commander of the Trans-Mississippi Department, Gen. E. Kirby Smith, may have been toying with the idea of seizing Close's Galveston foundry. Several reports had disclosed Close's lack of enthusiasm for the torpedo boat project, so General Smith (by way of his chief engineer) made inquiries in Houston concerning what course should be taken. In response to this inquiry from district headquarters, General Walker sent the following informative report regarding progress made:

> Houston, December 29, 1864. Captain W. A. Feret, Acting Chief of Engineers Department of the Trans-Mississippi. Sir: I have the honor to acknowledge the receipt of your communication of the 13th instant, and to inform you that the work upon the torpedo boat is progressing satisfactory, rendering it, I think, unnecessary to resort to the impressment of Mr. Close's Foundry, as he is now engaged, I believe, in good faith in the fabrication of the necessary machinery.
>
> Instead of building a complete engine for the boat, I have taken an old locomotive engine which will answer the purpose well, and is unfit for Rail Road purposes. The engine belongs to the branch road from Beaumont to Sabine Pass, and is, I am told, the property of alien enemies. There is one part connected with the construction of this boat that I desire to call to your attention to. You are aware that it is going on under the contract made by Captain Dunn with Mr. Chubb, which stipulates the payment of forty thousand dollars, when the work has been half completed.
>
> Unless this contract is set aside, it is due to the contractor that the first payment should be made now as more than half of his part of the work is al-

ready completed. I am satisfied that it will not be in the interest of the government to set aside the contract, to say nothing of the bad faith it would exhibit towards the contractor who has faithfully complied with his obligations to the government. It is due to Mr. Chubb the contractor, to state that if the boat is ever completed, and put in successful operation it will be due to the energy of Mr. Chubb who seems devoted heart and soul to the completion of the work. Signed, General Walker.[65]

General Walker must have considered the work to be progressing at a favorable pace since he recommended payment be made. With the procurement of a locomotive steam engine at Columbia some days earlier, General Walker seemed optimistic that the vessel might soon be in service, for he stated in his report that "we will save at least two months in completing the boat."[66]

During the final days of December 1864, while work on the ironclad torpedo boats proceeded under the watchful eyes of David Bradbury and Robert W. Dunn, in Union-controlled Little Rock, Arkansas, M. P. Hunnicutt, a Federal spy, was preparing to journey into what was left of Confederate territories west of the Mississippi.[67] For some two months Hunnicutt would masquerade as a businessman and travel throughout Confederate-controlled Louisiana and Texas, gathering information on troop movements, fortifications, and military morale. Hunnicutt was to make his way to occupied New Orleans by mid-March and report his findings to the Office of the Chief Signal Officer. Although most of what Hunnicutt discovered was of little value to the Union war effort (his official report was not filed in New Orleans until the closing weeks of the war), one piece of intelligence stumbled upon quite by accident, while traveling in a dusty stage across the Texas prairie, revealed information on perhaps the most daring operation ever hatched by members of Captain Singer's Secret Service Corps.

As Hunnicutt was leaving Little Rock on his mission into rebel territories, what was left of the eastern Confederacy was teetering on the brink of collapse. In the eastern theater the Shenandoah Valley had been destroyed as a granary for southern armies during the summer of 1864, which allowed General Grant to focus all his attention on what remained of Robert E. Lee's entrenched forces at Petersburg. In the western theater, the vital Georgia cities of Atlanta and Savannah had both fallen to General Sherman's forces, while John Bell Hood's campaign to reoccupy Tennessee had ended in the disastrous rout of his army at Nashville in mid-December 1864. While Confederate armies were giving up large chunks of territory, inflation was strangling what was left of the southern economy, and all hope of foreign intervention had long since vanished.

Federal armies were poised to slice through what remained of the Confederacy in the coming months, for General Sherman and his army were then comfortably in Savannah receiving supplies for a planned northward push through the Carolinas as soon as the weather permitted. Wilmington, North Carolina, the last viable Atlantic blockade-running port (Charleston had been effectively sealed by early 1865), was still open to traffic, but the Federals were then preparing to launch the largest combined operation of the war to seize it. Desertions plagued both armies, but southern soldiers, many of whom had families then living behind enemy lines, were especially hard hit, for communication with loved ones then residing in enemy territory was impossible.

It was perhaps this feeling of hopelessness that drove two soldiers attached to the Confederate schooner *Lecompte*, a small Federal vessel that had been captured at the Battle of Galveston two years earlier,[68] at anchor in Galveston Bay, to steal a small skiff and desert to the Union fleet. During the early-morning hours of January 2, 1865, Privates Jones and Hassenger surrendered themselves to the officer of the deck aboard the USS *Gertrude*. When questioned about harbor defenses and blockade-runner traffic, the two men not only named all vessels then in the harbor but also gave a good description of one of the torpedo boats then being constructed by members of Captain Singer's Secret Service Corps. A report was filed several hours after the men had surrendered:

> USS Gertrude off Galveston, Texas. January 2, 1865. Sir: This morning at 1:45 a boat came along containing two deserters from the rebel army, Henry Jones of Bray's Regiment of Cavalry and C. M. Hassenger of the same regiment. Both of these men have been attached to the rebel schooner Lecompte. They left the wharf at Galveston at eleven o'clock last night. There are inside four steam blockade runners and one large schooner name unknown. The schooner ashore upon the point is the "Lodo" and was inward bound. Her cargo has been taken out and the vessel abandoned. The enemy are building a torpedo boat at Goose Creek, One hundred and Forty feet long to be clad in rail road iron. Very respectfully, your obedient servant. Henry C. Wade, Acting Master Commanding.[69]

A follow-up report on the two deserters regarding additional testimony has not come to light, so we do not know how threatening Federal naval officers considered the ironclad torpedo boat. We know that J. D. Braman was quickly named superintendent of construction on the torpedo boat project on Buffalo Bayou.[70]

While construction on the torpedo boats continued, blockade-runner traffic in the Gulf of Mexico was increasing. With Charleston harbor having been effectively sealed, and the waters off Wilmington then swarming with Federal warships poised for a final assault on Fort Fisher (which guarded the entrance to the harbor), blockade-runner crews, after learning these facts, either turned back to their port of origin or changed course for the distant Confederate port at Galveston. A Union surgeon attached to a Federal vessel anchored off Galveston during the early months of 1865 described after the war what he remembered of his encounters with rebel blockade runners attempting to steam past the fleet into the Gulf of Mexico under cover of darkness:

> When a blockade runner got in and unloaded, her cargo of cotton was always ready, and instantly loaded, ready for a start. Then came a period of waiting for a favorable chance to run out—that meaning a stormy, dark night, when the low hulls of the vessels, painted a dirty white, were quite invisible a hundred yards away. At the last moment before dark, the bearings of each man-of-war outside would be carefully taken, and steam got up. This part we could see from out stations, and always had plenty of warning of a coming attempt. As soon as it was dark they would creep slowly down the channel, over the bar, and then, with every possible pound of steam and the greatest speed, would make a dash for our line. All we could hear was the beat of paddles upon the water—the sound in darkness is so deceptive that no one can tell from which direction it comes, and as nothing could be seen, we usually kept perfectly still and let them go. Indeed, at the speed with which they were going, even if we had seen them, only a shot could have overhauled them—our clumsy blockaders, never.[71]

It seems that capturing incoming vessels loaded with arms and ammunition had more priority to the Union war effort than stopping ships laden with bales of cotton from getting out. While blockade-runner traffic increased in the Gulf of Mexico, Baxter Watson, then on duty in Mobile with his old friend James McClintock, appears to have become somewhat impatient about raising funds for his electrically powered submarine. Watson wanted to start construction on the new submersible as soon as possible, as noted in his letter to General Beauregard:

> Mobile, Alabama. January 6, 1865. General P. G. T. Beauregard. Sir: Being the inventor of the submarine boat that destroyed the "Housatonic" in February last and losing all I had with her. I have concluded to lay the matter before you, and request your assistance in building another for the same place. . . . That boat as you know was a complete success as far as the boat was concerned. But was a complete failure through mismanagement, a fault over which I had no control as my supervision of her ceased as soon as she was in the water. . . . I propose to build another provided I can get the necessary assistance to do so. . . . It will require an electromagnetic engine to propel a boat of that description as a boat of that kind is impracticable with any other kind of power, but with it can be managed safely and successfully. . . . I firmly believe that I

can destroy the blockade in Charleston in a short time if I get the assistance. Your obedient servant, Baxter Watson.[72]

This letter is the last known wartime communication attributed to Baxter Watson, and nothing more regarding his plans to construct an electrically powered submarine has thus far come to light. With the war then winding down, Watson's unique proposal to build another submarine boat was probably known to only a handful of Captain Singer's group, for by early 1865, Singer personnel were spread throughout what remained of the crumbling Confederacy and in all likelihood had lost almost all contact with one another.

During this period Watson anxiously awaited replies to his pleas for funding to construct an electrically powered replacement for the *Hunley* (we do not know whom else besides Davis and Beauregard he may have contacted for funding), Fort Fisher, guarding Wilmington at the mouth of the Cape Fear River, fell to Federal forces.[73] On January 15, 1865, a combined Union attack from both land and sea (the largest of its kind ever launched during the Civil War) forced the Confederates defending the fort to surrender to an overwhelming force. With the collapse of Fort Fisher only the heavily blockaded Atlantic port of Charleston remained open to blockade-runner traffic, but by mid-January Confederates were already making plans to abandon the city because Sherman's resupplied army was then marching northward from Savannah.[74]

By mid-January 1865, while Union military strategists planned the upcoming spring campaign that would crush what remained of the eastern Confederacy, M. P. Hunnicutt was deep within Confederate-controlled territory. His journey had been plagued with misfortune from the start. Within days of entering regions still controlled by the rebel government, he had been robbed of everything he owned and locked up by General Magruder for being a suspected spy. He filed an official report in New Orleans in mid-March:

> I left Little Rock on the night of the 27th of December for Shreveport. I was robbed by Gus Crawford's Guerrillas of my pony, pistol, money and most of my clothing. I proceeded on foot and about four miles further met a flag of truce (Rebel) under the command of Lt. Col. John D. Bull. . . . Col. Bull gave me a pass to Washington, Arkansas to report to the Provost Marshal General. I reported to the Provost Marshal on the 6th of January 1865, and was paroled to report to the Provost Marshal General of the 2nd District of Arkansas daily at 9 a.m. On the 20th of January I was placed in the Guard House by order of General Magruder charged with being a confederate of a man by the name of Marston, alias C. S. Bell, whom was charged with being a spy.
>
> Bell's trial was going on when I left [Hunnicutt never stated the reason why General Magruder had released him], and from what I knew myself the evidence was very conclusive against him. I later heard . . . at Marshall Texas,

that Bell was convicted.... Magruder has issued orders for all irregular troops to join some regular organization. South of the Arkansas River the rolls show about twenty-five thousand regular troops in the state.... Magruder is fortifying on the Little Missouri, he is drunk about all the time.... The general impression of both soldiers and citizens is Arkansas will have to be given up although Kirby Smith has prohibited any more immigration to Texas from Arkansas particularly to parties taking Negroes....

I left Shreveport on the 2nd of February for Austin, via Marshall. Between Marshall and Shreveport there is sixteen miles of miserable railroad with one locomotive which is run in connection with the stages, connecting at Greenwood for Shreveport....

On the stage from Marshall Texas I became acquainted with a man by the name of James Jones who had been sent from Richmond Virginia, where he had been engaged in the torpedo business to destroy the Ram Tennessee. I saw his plans and specifications for constructing a boat 40 feet long, 40 inches deep and 48 inches wide for the purpose of destroying that vessel particularly. The hull of the vessel is already at Houston and there are four others at Shreveport.

I paid particular attention to his statements and agreed to go into the undertaking with him and invest five thousand dollars in the torpedo association if I did not go to the interior of Mexico. I wrote him from Bagdat that I would go into the interior. The boat is to leave Houston down the Brazos, thence into the Atchafalaya Bayou and through some other Bayou to the Red River. The reason that they did not take one of the boats from Shreveport was because there was so many Yankee spies around there he was afraid the plot would be discovered. They contemplate starting by the tenth of March and certainly by the twentieth. From all the circumstances connected, I consider that the attempt will be certain.[75]

The importance of the information revealed in the last few paragraphs of the report is irreplaceable to the history of the Singer Secret Service Corps, for no other document thus far discovered even hints that the Singer group had secretly built another submarine at Houston during the closing months of the war. The odds against a bumbling Union spy accidentally stumbling upon such a plot are overwhelming. If Hunnicutt had taken the previous day's stage or had never struck up a conversation with James Jones, we would never know of this bold Singer plan to destroy the ironclad *Tennessee* by a submarine attack.

From the way Hunnicutt described his accidental encounter with Jones, we can assume that the talkative Port Lavaca jeweler had no idea that he was giving away valuable information to an enemy spy, for Hunnicutt even stated that Jones had invited him to invest in the torpedo association of which Jones was a member. Hunnicutt stated in a later report that the submarines built by the torpedo association to which Jones was affiliated were "usually worked, seven feet underwater" and could be "propelled at the rate of four miles per hour, by means of a crank." Hunnicutt also stated in his final report that Jones and his associates had "also constructed the one which attempted to destroy the New Ironsides in Charleston" (obviously referring to the *Hunley*).[76] The information revealed in Hunnicutt's final debriefing taken in New Orleans seems to have caused a minor panic, for within hours after transcribing his testimony, a warning was immediately issued to the officers of the Mississippi Squadron to keep a sharp lookout for rebel submarine boats in the Red River.

At the same time that M. P. Hunnicutt was traveling southward toward his accidental encounter with James Jones, Maj. Henry Leovy and the men under his command had been busy rounding up various ring leaders of the treasonous order known as the Heroes of America. By early December 1864, at least ten assumed leaders of the order had been arrested throughout counties in southwestern Virginia and thrown in jail to await trial.[77] In years past the accused men may well have faced the gallows; however, times had changed by the winter of 1865, and what little remained of the crumbling Confederacy was quickly unraveling.

Among both civilians and soldiers, lawlessness had noticeably increased during the early weeks of 1865, and numerous bands of deserters freely roamed the Virginia countryside, plundering farms and property of helpless citizens. Home Guard units and other local organizations that had been delegated the responsibility of maintaining law and order were breaking down at an alarming rate, and to have men locked up during such times on charges of "disloyalty and harboring deserters" must have seemed counterproductive to Major Leovy and his subordinates. By early February 1865, most of the alleged leaders of the Heroes of America had been incarcerated in Major Leovy's crowded jail for nearly two months, and it seems that he was then considering letting them all go.[78]

We know so little about the fates of the incarcerated men because Jefferson Davis had promoted Leovy to colonel of cavalry in early February 1865 and given him a new command within days.[79] Within two months after receiving his new rank he would be in command of a cavalry detachment that accompanied President Davis and what then remained of the Confederate government on their desperate escape southward.

During this time, construction continued on the ironclad torpedo boats for the defense of Galveston and the Texas coast. The military was giving top priority to protection of the vessel being built at Captain Chubb's shipyard:

> Houston. February 21, 1865. Captain Thomas Chubb, Superintendent Steam boat repairs. Sir: You are instructed as follows: You will take charge of the

dredge boat and at once proceed to dig out the mouth of Tyber Creek and canal to a depth sufficient to float vessels drawing five feet nine inches water. If this can be done before the machinery is put into the torpedo boat, the later will not be removed: If this cannot be done in time, you will remove the torpedo boat to the ship yard at Lynchburg, and hasten forward the work as rapidly as possible.

If possible I desire that double reliefs be put to work on the torpedo boat, so as to carry it on night and day to completion. The importance of this cannot be over estimated and I feel assured you and Captain Bradbury will spare nothing in meeting my expectations and the aspirations of the country. The object of removing the torpedo boat to Lynchburg being to save it from being shut up in Tyber Creek, where it might be lost in case the enemy should get into the bay. Should you find that it will be more convenient to do the work at Spellman's Island, you are authorized to go to work at that point instead of at Lynchburg. G. Walker, Major General Commanding.[80]

Judging from the contents of this document, General Walker expected great things of the ironclad monster under construction and he wanted it completed as soon as possible. With news from incoming blockade-runner crews of the loss of both Wilmington and Charleston, it was logical to assume that additional Federal warships would soon be seen at anchor off Galveston.

While construction on the torpedo boats continued, Robert Dunn seems to have concluded that he needed additional torpedoes for the defense of his hometown of Port Lavaca. A Texas order book dating from this period states that he requested thirty-eight Singer mines be sent southward to obstruct the waterways leading into Calhoun County. With activities along the Texas coast expected to heat up in the coming weeks, Dunn apparently had taken it upon himself to protect the homes of his fellow Masonic Lodge members by deploying a deadly ring of submerged explosives.[81]

John Fretwell, then on duty in Virginia, was about to be sent once again into the wilds of North Carolina. With General Sherman's resupplied army moving northward from the smoldering ruins of South Carolina's capital, General Beauregard sent the following telegram to Gen. Braxton Bragg from his headquarters at Charlotte, North Carolina: "March 2, 1865. General Braxton Bragg. Sir: The movement of Sherman indicates with certainty an advance on Cheray; thence he will doubtless move forward upon Fayetteville. P. G. T. Beauregard."[82]

Since leaving the occupied city of Savannah some weeks earlier, General Sherman and his army, numbering about sixty thousand, were once again isolated in enemy territory with no means of communicating with the Federal War Department in Washington. Sherman had been in a similar situation several months earlier after putting Atlanta to the torch and starting his now legendary march to the sea. Cutting a path of destruction sixty miles wide across Georgia, Sherman's "Bummers," as they came to be known, had been out of communication for many weeks, and the only way officials at the Union War Department knew what progress was being made was by reading southern newspapers taken from captured Confederate soldiers.

But this crude form of communicating with Washington had been discovered, for on the same day that Beauregard sent the telegram warning of an advance on Fayetteville, Gen. Braxton Bragg sent the following to General Lee: "Goldsborough, March 2, 1865. General R. E. Lee, Richmond, Va. Sir: I find Colonel Childs has allowed a press dispatch to go from Fayetteville which gives Sherman's position, and will do us injury. I suggest the Richmond and Petersburg papers be not allowed to publish it. Braxton Bragg."[83]

The Confederates quickly realized that Sherman's army could be resupplied only if he could reach Fayetteville and the Cape Fear River. By ferrying barges up the Cape Fear from Wilmington, Sherman could obtain necessary supplies for a continued campaign through North Carolina and eventual march into Virginia. A telegram was quickly sent to Fretwell, who had halted a similar advance on the Roanoke River some three months earlier: "Goldsborough, March 2, 1865. Doctor Fretwell, Care of General Rains, Richmond: Wanted here immediately with all torpedoes on hand. No boats can be procured here. L. S. Baker, Brigadier-General." Within hours a message was sent to General Hoke, then garrisoned just south of Fayetteville on a small tributary that fed into the Cape Fear River: "Goldsborough, March 2, 1865. General Hoke, Rockfish Creek: Torpedo man in Richmond has been telegraphed for. Archer Anderson, Assistant Adjutant-General."[84]

On March 7, John Fretwell and his small staff of Singer operatives were en route south "with all torpedoes on hand,"[85] and General Sherman's veteran army crossed the Pee Dee River into North Carolina. From his headquarters at Laurel Hill Presbyterian Church, Sherman sent a detachment of cavalry to Wilmington with orders to send supplies up the Cape Fear River to meet him at his next destination, Fayetteville.[86] Although Sherman had issued orders stating that "every effort will be made to prevent any wanton destruction of property, or any unkind treatment of citizens," a North Carolina woman indicates that his directive, in many cases, fell on deaf ears: "There was no place, no chamber, trunk, drawer, desk, garret, closet or cellar that was private to their unholy eyes. Their rude hands spared nothing but our lives. Squad after squad unceasingly came and went and tramped through halls and rooms of our house day and night. At our house they killed every chicken, goose, turkey, cow, calf and every living thing, even our pet dog. They took from old men, women and children alike, every garment of wearing apparel save what we had on. Such as it did not suit them to take away they tore to pieces before our eyes."[87]

We can assume that Rockfish Creek and the Cape Fear River were to be Fretwell's theater of operations. With Sherman advancing on the all-important manufacturing center of Fayetteville, Fretwell and his subordinates were most likely given the task of mining the water approaches to the city prior to Sherman's expected arrival. Gen. Gabriel Rains gave the following account regarding Fretwell's late-war activities: "My sensitive primers were used by Dr. Fretwell in preference of all others. . . . I sent Dr. Fretwell down from Richmond, he arrived in time and according to my directions he succeeded in planting the torpedoes in the river above the enemy."[88] Unfortunately for the Confederates, Fretwell's submerged mines had little effect in slowing Sherman's advance, for on March 11 the city of Fayetteville was occupied.

While Fretwell and his operatives presumably played a cat-and-mouse game with scouting parties of Federal cavalry in central North Carolina, in what was left of Confederate-controlled Louisiana, James Jones was overseeing a similar operation along the Red River. On March 7, 1865, while Fretwell and his men were mining the waterways near Fayetteville, Ike Hutchinson and James Jones hatched a scheme to destroy the powerful 578-ton Federal ironclad USS *Ozark* then on station in the Red River. By attaching a single Singer torpedo at the center of a long buoyed rope stretched across the river, they hoped that the untested configuration, when set adrift in the river's current, might entangle itself around the *Ozark*'s hull, but their plan ended in failure when the buoyed line attached to the torpedo broke or snagged before reaching the anchored ironclad.[89] Since their new torpedo configuration was unsuited for the task, they presumably returned to traditional Singer tactics and oversaw the resumed anchoring of torpedoes in the Red River south of the Confederate Navy Yards at Shreveport.

While Hutchinson and Jones continued to monitor torpedo operations along the Red River in a desperate attempt to keep the growing Federal Mississippi Squadron at bay, M. P. Hunnicutt was then en route to occupied New Orleans. After parting ways with Jones in early February (and informing him by letter some days later that he was going to Mexico and would therefore be unable to invest five thousand dollars in the "torpedo association"), Hunnicutt continued by stage to the state capital at Austin and then on to San Antonio, where he continued to gather military intelligence.[90]

With his luck holding out, Hunnicutt continued south to Brownsville, Texas, gathering additional information on Confederate troop movements along the way, and then crossed the Rio Grande into Mexico on March 1, 1865. Hunnicutt immediately reported his findings to Major Jones, a Federal intelligence officer on duty at Brazos, then hastily booked passage on a New Orleans–bound steamship.[91] Upon arrival in Louisiana several days later, Hunnicutt immediately arranged a meeting with the acting head of the

Office of the Chief Signal Officer, Military Division of West Mississippi, and filed his report. With Hunnicutt's testimony then in hand, Maj. A. M. Jackson sent the following report to Lt. Col. C. T. Christensen, assistant adjutant general:

> Headquarters. Military Division of West Mississippi, Office of the Chief Signal Officer, New Orleans, La., March 13, 1865. Colonel: I have the honor to submit to your consideration the following report of information received at this office this 13th day of March 1865. In a letter from Captain Collins, Confederate scout, to a person in this city, he states that he expects a visit about this time from one Ike Hutchinson, from Lavaca, Texas, who has charge of the torpedoes in Red River. This taken in connection with Mr. Hunnicutt's report of the designs of Jones (also from Lavaca), who was at Houston, Texas, January 12, to destroy the ironclad Tennessee and other gunboats at the mouth of the Red River, leads me to believe that there is some such plan afoot, of which the commanders of gunboats should be notified.
>
> The following is a description of the torpedo boats, one of which is at Houston and four at Shreveport: The boat is forty feet long, forty eight inches deep and forty inches wide, built entirely of iron, and shaped similar to a steam boiler. The ends are sharp pointed. On the sides are two iron flanges (called fins) for the purpose of raising and lowering the boat in the water. The boat is propelled at the rate of four miles per hour, by means of a crank worked by two men. The wheel is on the propeller principle.
>
> The boat is usually worked, seven feet underwater and has four dead lights for the purpose of steering or taking observations. Each boat carr[ies] two torpedoes, one at the bow attached to a pole twenty feet long, one on the stern fastened on a plank ten or twelve feet long. The explosive of the missile on the bow is caused by coming in contact with the object intended to destroy. The one on the stern on the plank, is intended to explode when the plank strikes the vessel. The air arrangements are so constructed as to retain sufficient air for four men at work and four idle two or three hours.
>
> The torpedoes are made of sheet iron three sixteenths of an inch thick and containing forty pounds of powder. The shape is something after the pattern of a wooden churn, and about twenty eight inches long. Jones, the originator and constructor of these boats, also constructed the one which attempted to destroy the New Ironsides in Charleston, S.C. Very respectfully, your obedient servant, A. M. Jackson, Major, Tenth U.S. Colored Heavy Artillery.[92]

Hunnicutt and Major Jackson may have verbally discussed the background and mission of James Jones in some depth, for the report states that Jones was from Port Lavaca and in Houston on January 12, two important scraps of information noticeably absent in Hunnicutt's original handwritten report (Major Jackson obtained the information on Ike Hutchinson from other sources).[93]

Hunnicutt had paid close attention to the many details Jones revealed; Hunnicutt remembered the dimensions of the submarines (which exactly match the length and width of the *Hunley*) and recalled the speed at which the vessels could be propelled and the length of time that the boats could remain submerged with an eight-man crew (all of which again exactly match the characteristics of the *Hunley*).

Admiral Porter had learned of the four torpedo boats at Shreveport several months earlier and had ordered the mouth of the Red River obstructed to keep them out of the Mississippi.[94] However, verifiable Confederate records on the subject of the five Tran-Mississippi submarine boats have not yet come to light.

The fact that the submarines may have been under construction, and perhaps completed, prior to Jones's return to Texas is reinforced in Hunnicutt's original report, for he states that Jones told him that "the hull of the vessel is already at Houston and there are four others at Shreveport,"[95] information revealed in early February, only about a month after Jones had returned to Texas. Federal authorities in New Orleans took Hunnicutt's report quite seriously, and a copy was quickly sent to the naval commander on duty at the mouth of the Red River.

While Union naval officers on duty at the mouth of the Red River received warning to be on the lookout for rebel "infernal machines," at Captain Chubb's Galveston shipyard, work on the ironclad torpedo boat continued day and night. Robert Dunn and David Bradbury were then preparing it for harbor trials: "Houston, Tuesday, March 24, 1865. Special Order 83. Captain H. R. Marks, Post Commanding Houston, will issue to Mr. D. Bradbury chief constructor torpedo boat, ten pounds of candles."[96] Since oil lanterns give off toxic fumes, their use in a confined area would have been impossible. The "ten pounds of candles" were being issued to Captain Bradbury for the sole purpose of illuminating the interior of the torpedo boat. For surely blacksmiths and carpenters assigned to the project's night shift would have used adjustable oil lanterns and torches (not small flickering candles).

The Singer group knew that the heavy boat would be a sluggish, low-silhouetted lumbering giant whose only hope of approaching an enemy vessel undetected would be (like the *Hunley*) under cover of darkness. The following report to Capt. B. F. Sands, commanding the Second Division West Gulf Squadron off Galveston, summarizes testimony concerning deployment of the vessel taken from several deserters:

> USS Cayuga, Off Galveston, Texas, March 27, 1865. Sir: The following is about the substance of the information obtained from the seven men picked up by this vessel on the 24th instant: They are from Colonel Joseph Cook's heavy artillery regiment, stationed at Virginia Point. . . . They think that the

number of guns at Galveston has not been diminished, but they have frequently changed their position. All the guns have been newly mounted, the carriages manufactured at the general ordnance depot at Houston. They know nothing definite about the torpedo boats, but have heard that such boats were being built on the San Jacinto River, at Lynchburg. One of the men saw what was shown to him as a torpedo boat lying in the main channel at Galveston. It was shaped like a box, with square corners, and was quite low in the water. He could not tell whether she was plated or not. Your obedient servant, Henry Wilson, Lieutenant-Commander.[97]

This testimony is rather compelling because the description of the Galveston torpedo boat seems to match the design specifications outlined in Robert Dunn's 1864 letter to General Magruder. From documentation discovered in an obscure record group titled "Unfiled Papers and Slips Belonging in Confederate Compiled Service Records" at the National Archives, Captain Edgar Singer was himself back in Texas in late March 1865. He may have returned in January considering that Fretwell seems to have been in charge of torpedo operations in what then remained of Virginia and North Carolina at the time.[98] Singer's name is absent from the sparse documentation that survives, and he may have been overseeing the entire torpedo boat operation and had little time to deal with trivial orders regarding the issuing of candles and the redeployment of blacksmiths and carpenters (duties apparently assigned to his subordinates Dunn and Braman).

The events now under discussion took place during the final weeks of the war. Although most of the Confederacy east of the Mississippi River was then in Union hands, the once important blockade-running port of Mobile, where James McClintock and Baxter Watson were still on duty, had not yet fallen to the Federals. Whether Watson was still seeking funds to build his electrically powered submarine during this time is unknown. However, since Mobile Bay was then teeming with Federal ironclads and warships, he and McClintock had probably scrapped the proposed project and were focusing their attention on nocturnal torpedo deployments at the mouth of the Blakely River.[99]

Since the loss of Mobile Bay several months earlier, Gen. Dabney Maury had been actively preparing for a final assault on the city itself. Every approach to Mobile was protected by obstacles, ditches, subterra shells, and heavy gun emplacements, and every anchorage and channel that fed into the bay was mined with Singer torpedoes bobbing just beneath the surface.[100] The main obstruction deterring a Federal attack was Spanish Fort, a massive earthen fortification lying on the eastern shore of Mobile Bay that commanded all approaches to the city with a battery of six heavy guns and a garrison then numbering about two thousand men.

By mid-March the Federals decided that a joint army and naval attack on Spanish Fort was the only course of action, so with troops marched in from occupied Pensacola, Florida, a final assault on the earthen fortification was set for the morning of March 28. The commander of the double-turreted ironclad *Milwaukee* recorded his recollections of his unexpected encounter with a submerged Singer torpedo on the day of the assault, undoubtedly one of many that had been anchored some nights earlier under the direction of James McClintock and Baxter Watson:

> USS Genesee, Mobile Bay, March 30, 1865. Sir: I take the earliest opportunity to make a report of the sinking of the U.S.S. Milwaukee, under my command, on the afternoon of the 28th instant. I had proceeded up the Blakely River in company with the U.S.S. Winnebago to within about 1 ½ miles of the lower fort on the left bank of the river for the purpose of shelling a rebel transport supposed to be carrying supplies to the fort; had succeeded in causing the steamer to retreat up river. . . .
>
> I had returned within about 200 yards of the U.S. ironclad Kickapoo, then lying at anchor, and supposed the danger from torpedoes had passed, as I was where our boats had been sweeping, and also exactly in the same place where the U.S. ironclad Winnebago had turned not ten minutes before, when I felt a shock and saw at once that a torpedo had exploded on the port side of the vessel, aft the after turret, and, as near as I could determine at the time, about 40 feet from the stern. . . . The stern of the vessel sunk in about three minutes, as near as I can judge, but the forward compartments did not fill for nearly an hour afterwards, giving the crew an opportunity to save most of their effects. I saw every man off the vessel, sending them to the Kickapoo. J. H. Gillis, Lieutenant-Commander, U.S. Navy.[101]

On the following afternoon the ironclad monitor *Osage* shared a similar fate along the same stretch of river, when it accidentally drifted into another hidden Singer mine anchored near the wreck of the *Milwaukee*.[102] The loss of the two powerful ironclads came as a great shock to the Federals, who thought that the Blakely had been cleared of torpedoes several days earlier. A Confederate naval officer wrote after the war about these two late war victories accomplished by James McClintock and Baxter Watson:[103] "The destruction of these heavy ships caused much exultation among the Confederates, which found expressions in salutes from Spanish Fort and the guns of the Nashville. To the federals the two disasters, one following the other so closely, were depressing, as they had swept the channel for torpedoes regularly, and had taken up 120 within a few days previously."[104]

Although the victories against the two ironclads in the Blakely River had been complete, their loss had little effect in stopping the Union assault on Mobile. Within days after celebrating their greatest accomplishment since the sinking of the *Tecumseh* several months earlier, the city of Mobile was ordered abandoned. With the city in panic the breech-loading cannon that ex-*Hunley* crewman Lt. William Alexander had been called back from Charleston to design, was hastily loaded aboard a crowded troop transport, while McClintock and Watson, in all likelihood, oversaw the stowing of what torpedoes then remained. After all weaponry and ordnance that could be hauled away was onboard, the crowded transports carrying the last remnants of the Mobile garrison slowly steamed upriver toward the interior of Alabama. For the two submarine designers who had overseen the construction of the *Pioneer*, *American Diver*, and *Hunley*, their four-year struggle for southern independence had unceremoniously come to an end.

Taking the Fight to the Enemy 6

WHILE THE RAGGED SOLDIERS of the Mobile garrison retreated inland with hopes of linking up with what remained of Gen. Richard Taylor's army, a similar evacuation was taking place in war-ravaged Virginia. In late March, at about the same time that Union forces were planning a combined assault on Spanish Fort, Gen. Robert E. Lee and his Army of Northern Virginia were facing a critical situation. After months of being forced to extend their lines, Lee's depleted forces were spread across a thirty-mile front facing a well-supplied enemy that outnumbered them nearly three to one.

At dawn on March 25, 1865, General Lee, in a desperate gamble to break the Union center, launched his last great assault of the war against Fort Stedman, a fortified earthwork not two hundred yards from the Confederate lines. In a scene that would be repeated countless times during the First World War, the soldiers of Lee's tattered regiments gave a high-pitched rebel yell and swarmed from their trenches into no-man's land. Although the Confederates were exposed to withering fire from the fort, the bold nature of the attack had taken the Federals completely by surprise, and the fortification was quickly overrun.[1] Unfortunately for Lee and his army, Grant soon rallied a massive counterassault, and within four hours after first going over the top, the battle-worn southerners had been beaten back to their own trenches.

On the final day of March 1865, General Lee found his position steadily becoming undefendable and quickly made plans to evacuate his army to the southwest for a link with General Johnston's forces in central North Carolina. At 10:40 a.m. on Sunday April 2, with Federal shells bursting all along his front, Lee hastily sent a telegram to the secretary of war advising him that Richmond should be evacuated immediately.[2] With the sound of distant cannon fire clearly heard in Richmond that morning, a messenger from the War Department delivered General Lee's dispatch to President Davis, who was sitting in his usual pew during church services at St. Paul's. Eyewitnesses report that Davis slowly rose and unsteadily made his way up the aisle in silence as uneasy whispers echoed through the crowded congregation.[3]

While several of Lee's regiments fought a delaying action to cover the bulk of the army as they withdrew from the muddy trenches around Petersburg, Jefferson Davis and his small staff made their way back to the executive mansion. Davis described what happened next: "I quietly rose and left the church, went to my office, assembled the heads of departments and gave the needful instructions for our removal that night."[4] As word spread through the city, children ran about with the news while concerned groups gathered on street corners and discussed the circumstances that had led them to that day; then, like confused wildlife before an oncoming brush fire, they began the evacuation of the Confederate capital.

"Suddenly as if by magic, the streets became filled with men, walking as though for a wager," wrote an eyewitness after the war, "and behind them excited Negroes with trunks, bundles and luggage of every description. All over the city it was the same—wagons, trunks, bandboxes and their owners, a mass of hurrying fugitives filling the streets."[5] Another witness to the chaos later wrote, "Thousands of citizens were determined to evacuate the city with the government. Vehicles commanded any price in any currency possessed by the individual desiring to escape from the doomed capital. The streets were filled with excited crowds scurrying to the different avenues for transportation, inter-mingled with porters carrying huge loads, and wagons piled up with heaps of baggage, of all sorts and descriptions."[6]

With instructions presumably received from the president, Secretary of State Judah P. Benjamin returned to his office at the Department of State, withdrew fifteen hundred dollars in gold from the Secret Service Fund, and then oversaw the destruction of all files concerning covert operations, personnel, and so forth that could be assembled in the brief time allotted.[7] His clerks scurried from office to office throughout the afternoon, hastily assembling boxloads of important documents and packaging them for transportation to the railroad station.

While Benjamin oversaw the destruction of all Secret Service–related documentation that could be assembled, President Davis was issuing special orders to the Treasury Department to box up all remaining gold and silver (a little over half a million dollars' worth) and transfer it to the railroad station to be put aboard a special train. With the Confederate Treasury secured and members of his cabinet and their staff en route to the train station, Davis ordered the executive mansion to be cleaned from top to bottom. As numerous servants dusted and polished what remained, President Davis slowly walked through the front door, mounted his horse Kentucky, and with a small staff of military officers in attendance, silently rode to the train station through the scenes of chaos that had begun to grip the city.[8]

With the city in near panic, it was decided that the vast military stores of liquor on hand should be poured into the street to keep it from Federal troops: "About dusk the government commissaries began the destruction of their immense quantities of stores. Several hundred soldiers and citizens gathered in front of the building, and contrived to catch most of the liquor in pitchers, bottles and basins, that was poured out. This liquor was not slow in manifesting itself. The crowd became a mob and began to howl. Soon other crowds had collected in front of other government warehouses. At some, attempts were made to distribute supplies, but so frenzied had the mob become, that the officers in charge, in many cases, had to flee for their lives."[9]

Secretary of State Judah Benjamin, Secretary of the Navy Stephen Mallory, Secretary of the Treasury George Trenholm (who according to one source had attempted to purchase the *Hunley* from the Singer group soon after its arrival in Charleston),[10] Postmaster General John Reagan, Attorney General George Davis, and their staffs boarded the trains that carried the Confederate Treasury and slowly left Richmond depot after dark. One train carried government officials while the other transported gold guarded by sixty midshipmen from the Confederate Naval Academy. When the train carrying the Confederate government had safely crossed the James River, orders were given to torch the trestle to block any pursuit.[11]

On the following morning Secretary of War John Breckinridge and his staff rode out of Richmond and proceeded to General Lee's headquarters.[12] While the capital of the Confederacy burned to their rear, Breckinridge and those who accompanied him made plans to join the president and his cabinet at the new seat of government in southern Virginia. Within hours after Breckinridge and his officers had retreated from the ruins of the collapsed capital, Federal troops occupied the city. By that afternoon the Stars and Stripes were once again flying from atop city hall.

President Davis arrived in Danville about three o'clock on the afternoon of April 3, 1865,[13] and immediately made plans to issue the following optimistic proclamation to the citizens of the Confederacy. Although somewhat rambling in its content, the document seems to have summed up well Davis's unrealistic opinions about continuing the war:

> To the People of the Confederate States of America: We have now entered a new phase of a struggle the memory of which is to endure for all ages. . . . Relieved from the necessity of guarding cities and particular points, important but not vital to your defense, with an army free to move from point to point and strike on detail detachments and garrisons of the enemy, operating in the interior of our own country, where supplies are more accessible, and where the foe will be far removed from his own base and cut off from all reserves, nothing is now needed to render our triumph certain but the exhibition of our own unquestionable resolve. . . . Let us not, despond, my countrymen, but relying on the never failing mercies and protecting care of our God, let us meet the foe with fresh defiance, with unconquered and unconquerable hearts. Jefferson Davis.[14]

While Jefferson Davis and his cabinet prepared to release the proclamation, along the windswept coast of Galveston Bay, Capt. Edgar Singer and his operatives appear to have shared a different impression of the conflict, for in many ways they had not yet begun to fight. On the morning of April 4, 1865, the following order was issued in Houston: "Special Order Number 76 ... Captain Stanfer's Company of Unattached Cavalry will report with as many of his men as may be necessary to Mr. R. W. Dunn for service on the torpedo boats now being constructed in the district. Such members of his command as may not be required in the torpedo service will within ten days from reception of this order, join some infantry or artillery company in this district, as they may elect."[15]

It was not uncommon to temporarily detach troopers from various Texas cavalry regiments to act as crewmen aboard Galveston-based warships during times of crisis, such as the "Horsemen of the Sea" at the Battle of Galveston. We can assume that several such troopers attached to Captain Stanfer's Company of Unattached Cavalry were being reassigned as crewmen aboard the ironclad torpedo boat being constructed at Captain Chubb's shipyard.

One of the Singer ironclads was being prepared for deployment, according to a confidential report to Naval Secretary Stephen Mallory in Richmond, on April 8, 1865: "I have ordered Lieutenant Phillips to Texas to take command of a torpedo boat built near Galveston by the Engineer Department. So far as I can ascertain no demonstrations are being made by the enemy in the Trans-Mississippi Department. Your Obedient Servant, J. H. Carter, Commanding Naval Defenses Western Louisiana."[16]

We can conclude that on April 8, troops west of the Mississippi River did not know about the fall of Richmond, so Secretary Mallory (then with Jefferson Davis in Danville, Virginia) probably never saw Carter's report or received notice that Captain Singer's torpedo boats were nearing completion. After the report was sent (by an unknown route), Gen. E. Kirby Smith endorsed Carter's decision to dispatch Lieutenant Phillips to Houston and issued the following order: "Headquarters Trans-Mississippi Department Shreveport, La. April 10, 1865. Special Order No. 85. Lieutenant J. L. Phillips, C. S. Navy will report to Major General Magruder, Commanding District of Texas to be assigned to the command of the Torpedo Boat, now being constructed at Buffalo Bayou. He will report to Colonel Douglas, Chief Engineer Trans-Mississippi Department for instructions. By Command of General E. Kirby Smith."[17]

Apparently, officers attached to General Smith's Trans-Mississippi Department during the spring of 1865 had no intention of giving up the fight. Although the Confederate states east of the Mississippi then lay in ruin, Union forces had not touched most of the Indian territories (under command of the Native American Confederate general Stand Watie), Texas, and much of western Arkansas and Louisiana, and their attempts to enter these regions had been decisively repelled. A *New York Times* editorial discusses conditions in the Trans-Mississippi theater:

> Of all the states of the Southern Confederacy, the State of Texas has suffered the least—immeasurably the least—by the war. . . . Its comparative security has attracted to it tens of thousands of settlers from other parts of the south—from Louisiana, Alabama, Missouri and also from the Mississippi States. Its population, by the last census, was 600,000; but there is reason to believe that it now counts over 1,000,000, about three-fourths of whom are white. Though its ports have been blockaded, neither its agricultural nor its general resources have materially suffered. Its crops of cereal have been good, and its cotton crop larger than ever. For a great part of its cotton crop it has found an outlet by way of Mexico; and the large price in coin commanded by this article has made money abundant and general supplies plentiful.[18]

While Confederate naval officer Lt. J. L. Phillips made plans to journey from Shreveport to Houston to take command of one of the Singer torpedo boats in Virginia, General Lee and his hopelessly outnumbered army were preparing for the end. With all southward avenues of escape then blocked, Lee sent word to General Grant that he was ready to discuss terms of surrender. On the morning of April 9, 1865, at the tiny settlement of Appomattox Courthouse, Gen. Robert E. Lee surrendered what remained of his Army of Northern Virginia to Gen. Ulysses S. Grant. With the capitulation of Lee's army, Confederate resistance in the state of Virginia effectively came to an end.

On the afternoon of April 10, official news of Lee's surrender reached President Davis and his cabinet officers in Danville. "This news," wrote Naval Secretary Stephen Mallory some years later, "fell upon the ears of all like a fireball in the night."[19] The gold from the Confederate Treasury was already en route to Charlotte, North Carolina,[20] so Davis, his cabinet, and a multitude of military officers prepared to retreat southward to Greensboro. Traveling through the night, the train carrying the Confederate government arrived at Greensboro depot the following day, only to find the population of the city despondent over news of Lee's surrender.

On the following morning Generals Johnston and Beauregard met with President Davis and his cabinet and presented their views about continuation of the conflict. Although both Jefferson Davis and Secretary of State Judah Benjamin were of the opinion that the South should fight on, General Johnston, then facing a veteran Federal army under Sherman that outnumbered his nearly three to one, told the president and his cabinet officers that surrender terms should be negotiated.[21] With great reluctance President Davis granted Johnston's request to enter into negotiations with General Sherman (which he believed would fail) and then immediately made plans to continue moving the government southward.

As members of the government continued by wagon toward Charlotte, the governor of North Carolina, Zebulon Vance, reached Greensboro from the state capital at Raleigh too late to join the retreating government. Because Federals had overrun his state and General Johnston was preparing to negotiate a surrender with General Sherman, Vance was eager to meet with Davis to discuss future actions. When Governor Vance and his small escort finally overtook the retreating wagons, Davis immediately met with him to discuss the current situation.[22]

President Davis appeared optimistic, Vance later wrote, and "told me of the possibility, as he thought, of retreating beyond the Mississippi with large sections of the soldiers still faithful to the Confederate cause, and resuming operations with General Kirby Smith's forces as a nucleus in those distant regions." Vance went on to say that the president expressed "a desire that I should accompany him, with such of the North Carolina troops as I might be able to influence to that end. He was very earnest and displayed a remarkable knowledge of the opinions and resources of the people of the Confederacy, as well as a most dauntless sprit." The governor later said that there was a "sad silence" around the council table for several moments following the president's remarks; then one by one several cabinet members came out in support of Davis's plan.[23]

While Governor Vance was left to deal with the situation in North Carolina, Jefferson Davis and his administration continued their flight with hopes of ultimately establishing a new Confederate capital in the fertile regions west of the Mississippi River. Determined not to give up the struggle for independence, Davis believed that he could gather up loose ends of his shattered government and disorganized military units in the east, force a crossing of the Mississippi, and establish a western Confederacy supported and defended by cotton. As Davis and his followers continued their retreat through North Carolina, they unanimously decided that Texas would be their final objective. This determination to reach the Lone Star State was intensified by the constant danger of capture, which increased every day they remained east of the Mississippi.[24] Desertions increased among the accompanying cavalry troopers who were well aware of the fact that most of the eastern Confederacy had already collapsed and that homeward-bound paroled soldiers from Lee's army were then roaming the countryside.

Since Texas was free of Federal forces and would probably remain so for at least the foreseeable future, President Davis probably viewed that territory as the last defendable region of the Confederacy. Galveston was still open to blockade-runner traffic and military supplies were readily available from just across the Rio Grande in friendly Mexico, so organized resistance to Federal authority could perhaps be prolonged indefinitely. Confederate general John Magruder stated on April 28, 1865, that sufficient Federal forces capable of invading Texas could not be mounted until the spring of 1866.[25] Davis and his cabinet, meanwhile, were completely unaware of events then taking place that would quickly cause many in the North to scream for their blood.

On the evening of April 14, 1865, while attending a play at Ford's Theater with his wife, President Abraham Lincoln was assassinated by the well-known actor John Wilkes Booth. The death of Lincoln had grave consequences for Davis and those who traveled with him, because within hours of Lincoln's death, word that the leaders of the rebel government had in some way been responsible for this act of barbarism filled the editorial columns of newspapers throughout the land.

President Davis and his cabinet received word of Lincoln's death soon after their arrival in Charlotte. After discussing his views regarding the continuation of the struggle with a regiment of North Carolina cavalry who had cheered him into giving a speech, President Davis was handed a telegram containing news of the assassination. Davis's initial reaction to the news was that Lincoln's death would be catastrophic to the citizens of the Confederacy already under occupation, for it was well known that Vice President Andrew Johnson did not share Lincoln's compassionate views about how to deal with a defeated South.[26]

Perhaps the man most distraught over the news of Lincoln's death was Judah Benjamin, the Confederate secretary of state whose duties also extended to both overseeing and funding operations of the Secret Service.[27] When news reached Benjamin that northern newspapers were blaming the fleeing Confederate leadership for masterminding the assassination, he must have been deeply shaken, for he above all others in the government could be linked to covert activities that had taken place behind Union lines. To complicate matters further (unbeknown to Benjamin at the time), soon after the assassination cipher codes were discovered in one of Booth's abandoned trunks at the National Hotel that were identical to Confederate codes discovered in Benjamin's State Department office at Richmond.[28]

With Lincoln then dead and the northern press howling for the head of Jefferson Davis and his assumed co-conspirators, escape to Gen. Kirby Smith's Trans-Mississippi Department took on a new urgency. While Federal troops combed the southern states for any sign of President Davis and his cabinet, Confederate general John Echols, accompanied by a detachment of cavalry under the command Col. Henry J. Leovy (who had been friends with Benjamin in New Orleans), arrived in Charlotte.[29] Whether Echols and Leovy had been under orders to follow what remained of the government into North Carolina is not known. Remaining documentation simply states that General Echols and Colonel Leovy were in Charlotte soon after the president's arrival. They may have been assigned the task as acting as a rear guard for the retreating government.

Not long after Henry Leovy's arrival and reacquaintance with his old friend Secretary of State Benjamin, plans were being hatched to send a squad of cavalry to Abbeville, South Carolina. Some days earlier Davis's wife, their four children, and the wives of several government officials had retreated from Charlotte to Abbeville with both the Confederate Treasury and the sixty midshipmen assigned to protect it. Henry Leovy's own wife was temporarily residing in Abbeville, so the colonel quickly volunteered to accompany the small detachment southward.[30]

While Colonel Leovy and a small squad of cavalry troopers made their way into South Carolina to find the president's wife and those who had accompanied her, Jefferson Davis was preparing to convene another cabinet meeting to debate the current situation. Davis and his officers mapped out strategies for the continued retreat of the government through South Carolina, Georgia, Mississippi, and across to Texas. At the conclusion of the meeting, Attorney General George Davis, weary of the situation, informed the president that he was resigning his post so he could care for his motherless children and, if at all possible, his personal property in Wilmington. On the afternoon of April 26, 1865 (the same day that General Johnston surrendered his army to Sherman, and John Wilkes Booth was cornered in a Virginia barn and killed by his Federal pursuers), the long government wagon train containing supplies, baggage, and the remnants of the Confederate archives slowly headed south from Charlotte.[31]

The wagons carrying the remnants of the fleeing Confederate government moved through a region of South Carolina that Sherman's army had bypassed. Because these scattered farms and plantations had never been touched by the war, the retreat through the towns that dotted the landscape in many ways resembled a triumphal tour, for wherever the presidential caravan went, cheering townspeople lined the streets to greet them as they passed. While traveling though this region, Secretary of the Treasury George Trenholm, who had been gravely ill since departing Richmond some weeks earlier, resigned his office (his official duties were quickly bestowed on Postmaster General John Reagan).[32]

Three days after President Davis and his escort had left Charlotte, Colonel Leovy and his small detachment of cavalry were preparing to link up with the government wagon train. Apparently there was some confusion about what westward route the government was planning to take:

> Abbeville, S.C., April 29, 1865—7:30 a.m. Mr. President: We had intended starting yesterday afternoon, but were detained by the rain. Are just about getting off now. The ladies and children are very well, and in good spirits. They

move in a good ambulance and carriage, and will reach Washington in a two days' drive from this place. From Washington we shall go toward Atlanta; there to halt, until we see or hear from you. This movement was determined by your telegrams, and by the belief that you would move westward, along a line running north of this place. Colonel Leovy has been kind enough to set out from here to meet you, to explain our plans, etc. He will tell you everything. With sincere prayers and hopes for your health and safety, very respectfully, your obedient servant, Burton N. Harrison.[33]

Henry Leovy was unaware that Jefferson Davis and his tattered escort were themselves heading toward Abbeville. Whether Colonel Leovy met with the president prior to his arrival in the city is not known. Once the Confederate government established official residence at Abbeville, Davis pursued with renewed vigor his vision of a new Confederacy rising from the ashes. With predictions that thousands would flock to the resurrected banner, President Davis and Judah Benjamin met privately on the morning of May 1 to discuss ways in which Confederate agents operating in Europe could be contacted to purchase munitions for the continuing struggle.[34] Benjamin wrote about his discussion with Davis: "I proposed to the president that, as we could not communicate with our agents abroad in any other way, I should leave him to pursue his journey across the country to the Trans-Mississippi and proceed myself to the Florida coast, cross to the islands, give the necessary orders and instructions to all our foreign agents, and rejoin him in Texas, via Matamoras [Mexico]. The plan was highly approved . . . with no one in on the secret of my purpose except the president and the cabinet."[35]

As Federal scouting parties searched for Davis and his administration, word of Lee's surrender reached General Smith in Shreveport.[36] His theater of operations was relatively free of Union forces and his army was well supplied with both food and munitions, so he defiantly issued a proclamation to his troops soon after the devastating news of Lee's capitulation had reached his headquarters:

> Headquarters Trans-Mississippi Department, Shreveport, La., April 21, 1865. Soldiers of the Trans-Mississippi Army: The crisis of our revolution is at hand. Great disasters have overtaken us. The Army of Northern Virginia and our Commander-in-Chief [General Lee] are prisoners of war. With you rests the hopes of our nation, and upon your action depends the fate of our people. I appeal to you in the name of the cause you have so heroically maintained—in the name of your firesides and families so dear to you—in the name of your bleeding country, whose future is in your hands. Show that you are worthy of your position in history. Prove to the world that your hearts have not failed in the hour of disaster, and that at the last moment you will sustain the holy cause which has been so gloriously battled for by your brethren east of the Mis-

sissippi. You possess the means of long-resisting invasion. . . . Stand by your colors—maintain your discipline. E. Kirby Smith, General.[37]

Forty-eight hours after issuing the proclamation to the Army of the Trans-Mississippi (which encompassed all military forces operating in regions west of the Mississippi River), Gen. John Magruder released a general order compelling the citizens and soldiers of Texas not to give up the fight and force the invading enemy to "pay dearly for every inch of territory he may acquire."[38]

From intelligence apparently gathered from Federal newspapers, it appears that the escape of Jefferson Davis and his cabinet was well known in regions west of the Mississippi River. With no other place for the Confederate government to retreat to, Kirby Smith and his staff officers at Shreveport realized the president's destination was their own theater of operations. Smith discussed plans to establish an army of fifteen thousand troops at Marshall, Texas, to be placed at the disposal of government officials upon their arrival.[39]

Rumors of imminent invasion circulated throughout regions west of the Mississippi, and Kirby Smith awaited arrival of President Davis, Lt. J. L. Phillips, who had been dispatched from Shreveport several days earlier, arrived in Houston to take command of one of Captain Singer's torpedo boats. But construction of the vessel had not been completed, prompting General Magruder's headquarters to issue the following order: "Houston, April 25, 1865. Special Order Number 115. Lieutenant J. L. Phillips of the C. S. Navy, will report to the Commanding Officer of Galveston, for temporary assignment to duty. Lieutenant Phillips will be relieved from such duty as soon as the Torpedo Boat, which it is intended he shall command, may be completed."[40]

General Magruder and his officers were then preparing to order the removal of several Galveston-based manufacturing facilities inland, as noted in a directive in a late war order book kept at Houston headquarters: "Houston, April 27, 1865. Major A. J. Lindsey, Superintendent of Foundries. Sir: The Major General Commanding desires you to have the Foundry at Galveston removed at the earliest practical moment. In making the removal however, you will let it be used as long as possible for the torpedo boat."[41] General Magruder most likely had high hopes for the huge secret weapon nearing completion at Chubb's shipyard. Troopers from Captain Stanfer's Company of Unattached Cavalry probably were being drilled in regard to their future duties, and we can presume that headquarters thought that it would be only a matter of days before the lumbering giant put to sea. A message from Robert Dunn to military headquarters reinforces that the vessel was being prepared for service:

Houston, Texas. April 28, 1865. Major General Magruder, Commanding the District of Texas, New Mexico and Arizona, General: I am informed by

Captain Carlin of Steamer Imogen that he brought in on his last trip some 200 tons of coal for account of Government. I respectfully request that a part of this coal be turned over to me for Torpedo Service.

I would also state that Captain Carlin brought in an Engine and Boiler complete for the construction of a Torpedo Boat. This Engine and Boiler was intended for Wilmington but in consequence of the occupation of that place by the Federals, was brought to Galveston. If you will turn this machinery over to the Torpedo Service a boat can be constructed at short notice under the direction of Mr. Bradbury and Captain William Lubbock. Very Respectfully Your Obedient Servant, R. W. Dunn of Singer's Special Service Corps.[42]

On the reverse of this letter General Magruder wrote: "Houston Texas, April 28, 1865. Mr. Sampson is requested to purchase the 200 tons of coal mentioned within, it is very greatly needed. It does not appear whether the 'Engine and Boiler' is the property of the South or not. Mr. Sampson is respectfully requested to ascertain if it is in my power to obtain these articles. J. Magruder, Major-General Commanding."[43]

Based upon General Magruder's endorsement, Robert Dunn and various members of Captain Singer's group may well have been given whatever articles they thought necessary for the defense of the Texas coast. Additional Federal blockading vessels were expected daily, so Sampson most likely negotiated a deal for the engine and boiler. It is highly doubtful that Carlin would have opted to rerun the blockade with such a useless, weighty item stowed below deck, since he would have wanted to pack his vessel with valuable cotton for a return trip. So Dunn more than likely succeeded in procuring these important items from which he planned to build a smaller torpedo boat.

Robert Dunn quickly turned his attention to preparing torpedoes for the nearly completed ironclad: "Houston. April 29, 1865. To E. Douglas Chief Engineer Trans-Mississippi District Shreveport, La. Sir: Would you please have the Treasury Department agent here purchase 585 pounds of copper to make torpedoes for use on boat. Signed R. W. Dunn esq. Troops Torpedo Service."[44] While Robert Dunn presumably prepared facilities at the Close Foundry to cast heavy copper torpedo casings, military authorities issued the following directive regarding the construction of the sister vessel then being built by Singer personnel near Houston at Lubbock's Mill:

> Headquarters District of Texas, Houston. April 30, 1865. Special Order 120. The torpedo boat in process of construction at Lubbock's Mill on Buffalo Bayou, under the superintendence of J. D. Braman of Singer's Special Service Corps., and commenced under order of Major Walker, will be constructed to completion. Captain Henry S. Lubbock, commanding Marine Department, will afford Mr. Braman, all the assistance in his power to facilitate the work, and which may not interfere with the operations of the Marine Department. The

Department of Texas, New Mexico and Arizona will fill all requisitions for materials for this boat, made by Lt. F. A. Rice A. A. Q. M. Marine Department. As the early completion of this boat is desired, Captain Lubbock will put upon the work as many mechanics as possible. Mr. Braman is instructed to make his reports through the commander of the Marine Department.[45]

While military headquarters pressured John Braman to bring his torpedo boat project to completion, Jefferson Davis and his administration were preparing to meet to discuss the deteriorating situation and take swift action before they found themselves cut off and trapped in the eastern regions of a dead Confederacy. On May 2, 1865, twenty-four hours after General Magruder had ordered the second Singer torpedo boat to be completed, Jefferson Davis called his cabinet together for the last time. "It was a historic scene," wrote a witness to the proceedings years later. "Mr. Davis presided, with General Bragg, who had become by the surrender of Lee, Johnston, Beauregard and Cooper, the senior general of the Confederacy, on his right hand, and General Breckinridge, Secretary of War and Major General, on the other side."[46] As the meeting progressed, a feeling of hopelessness descended over the large room where the group had assembled; one by one everyone present (except Secretary of State Benjamin and Davis himself) voiced the opinion that all was lost and that a call for continued resistance east of the Mississippi would be a cruel injustice to the people of the South.

At the conclusion of the meeting Secretary of the Navy Stephen Mallory resigned his post and informed the president that he would not be accompanying him to the Trans-Mississippi. With but three of his ministers left in attendance (Benjamin, Breckinridge, and Reagan), President Davis ordered the destruction of many official papers to lighten the load that would be transferred west. The most valuable documents from the Confederate archives were placed in the secret care of Col. Henry Leovy's wife prior to their departure, and the last remnants of the Richmond government, with a ragtag military escort in attendance, departed Abbeville on the afternoon of May 3.[47]

As they crossed the Savannah River into Georgia early the next day with hard riding ahead for the presidential party, the time had come for Secretary Benjamin to embark on his secret mission to contact Confederate agents abroad.[48] If all went well with the daring plan and he was successful in arranging the continued transport of arms to Texas, Benjamin would rejoin the transplanted government in the west in the months to come. In an old buggy pulled by two broken-down grays, Judah Benjamin and Henry Leovy split from the group on May 4, 1865, the same day that Gen. Richard Taylor (son of former president Zachary Taylor) surrendered the last Confederate army still in the field east of the Mississippi.[49] An eyewitness describes Benjamin at his departure : "With goggles on, his beard grown, a hat well over his face, and a large cloak hiding his features, no one would have recognized him as the late secretary of state of the Confederacy."[50]

Both Benjamin and Leovy were fluent in French, so they decided that Leovy would pose as a French interpreter, while Secretary Benjamin took on the guise of M. M. Bonfals, a Frenchman traveling the southern states on the lookout for cheap land.[51] Keeping to the back roads and avoiding towns as much as possible, Benjamin and Leovy continued southward toward Florida, hoping that no one would discover their true identities, since they were the last active operatives of the Confederate Secret Service. Their mission for the Confederate government could breathe new life into a defeated South. Because they had numerous accounts in both the Bahamas and Europe, Confederate agents operating abroad had ample funds from which to purchase munitions for the continuation of the struggle. If Benjamin and Leovy could only get word to them that President Davis was then planning to reestablish the government in Texas, agents could send huge shipments of arms and ammunition to Galveston.[52]

They shared the roads with returning veterans recently paroled from Lee, Johnston, and Taylor's armies and undoubtedly spoke French throughout their journey. If their true identities and mission were accidentally discovered, the two men could well face the gallows. In this manner Benjamin and Leovy traveled about thirty miles a day.[53]

In May 1900 Henry Leovy described their journey: "On the invitation of Mr. Benjamin I accompanied him on his escape.... Traveling in disguise, sleeping at night in log huts, living on the plainest fare, subjected to all the discomforts of such a journey, and with all his plans shattered and without definite hope for a future, his superb confidence and courage raised him above all, and he was the great, confident, cheerful leader that he had been in the days of highest prosperity."[54]

President Jefferson Davis and his shrinking entourage had decided to split up on May 6 near Sandersville, Georgia, with an understanding that the government caravan (then consisting of about seven wagons and five ambulances) would reunite near Tallahassee or Madison, Florida, in the days to come.[55] Accompanied by a small military escort, President Davis, with his wife and children comfortably situated in an ambulance, rode off to the southwest with hopes of rejoining the slow-moving, main baggage train in northern Florida. It was not to be.

On May 10, 1865, near the small Georgia town of Irwinville, President Jefferson Davis and a small entourage consisting of about forty individuals (ranging from family members to servants) were surprised and captured at dawn by a detachment of the 4th Michigan Cavalry who had been tracking them for several days.⁵⁶ Hope for a resurrected Confederacy rising from the ashes west of the Mississippi River dimmed.

While Davis and his aides surrendered what few arms remained, Col. John Taylor Wood, a trusted member of the president's entourage, gained his freedom by bribing a Federal cavalrymen (whom he had found rifling through the president's personal trunks), mounted an unattended horse, and galloped off to report the capture to Secretary of State Judah Benjamin. On May 11, just eight days after parting ways with the presidential caravan, and some twenty-four miles from Irwinville, Colonel Wood overtook Judah Benjamin and Henry Leovy and informed them of the president's capture.⁵⁷ Convinced that his own surrender would result in nothing less than a swift trip to the gallows (the common belief throughout the South was that all high-ranking Confederate officials, including Lee himself, would ultimately be hung for treason), Benjamin revised his plans and continued with Leovy toward Florida.⁵⁸

Soon after the two travelers had crossed into northern Florida, Secretary Benjamin decided he need to change his disguise. A farmer's wife helped him fashion some homespun clothes identical to those worn by her husband, and Benjamin and Leovy became South Carolina farmers in search of new land for their families.⁵⁹ Benjamin and Leovy undoubtedly discussed at length the logic of continuing their secret mission to contact southern purchasing agents abroad.⁶⁰ With Davis and his entourage then in Federal custody, the futility of continuing their struggle for independence must have become more evident with each passing mile.

Practically all members of the Confederate government were then under arrest, so Benjamin was painfully aware that his mission to contact purchasing agents abroad had become irrelevant and his only hope of avoiding the hangman's noose was to leave North America. To hasten his escape, and perhaps save his old friend Henry Leovy from being implicated as an accomplice, the two men parted company within days after receiving word of Davis's capture.⁶¹

As Judah Benjamin continued his flight down Florida's western coast, Henry Leovy made his way to Tallahassee and surrendered to Federal authorities on May 21, 1865.⁶² With Federal passes and a recently signed parole document in hand, Leovy made plans to return to his wife in Abbeville, where, unbeknown to Colonel Leovy's parole officers, she had been delegated the responsibility of secretly hiding a large portion of the Confederate archives.⁶³ We do not know when or how the Federal government eventually gained access to the archives or how long Leovy and his wife took it upon themselves to oversee the papers' safekeeping.⁶⁴

While all military actions in the eastern regions of the collapsed Confederacy gradually came to an end, Confederate naval officer J. L. Phillips was at last to be officially assigned to the ironclad torpedo boat then nearing completion. Although the enormous vessel was not quite ready for service, Lieutenant Phillips was to do all in his power to ensure its completion as soon as possible: "Houston. May 9, 1865. Special Order Number 129. In accordance with instructions from Department Headquarters, Lieutenant J. L. Phillips, C. S. Navy will take command of the Torpedo Boat, now being constructed on the Bayou, as soon as it is completed. He will in the meantime devote his attention to such arrangements and preparations as may be necessary to insure its rapid completion."[65]

The huge ironclad torpedo boats being built at Chubb's shipyard and Lubbock's Mill were unquestionably the last Confederate warships under construction on southern soil. General Magruder had purchased some two hundred tons of coal for torpedo service in late April, and presumably the locomotive boiler and engine were then in place aboard the torpedo boat and ready for service.

From various orders and dispatches transcribed from surviving Trans-Mississippi order books, May 9, 1865, was a busy day for Gen. Kirby Smith. He officially sent word to Gen. John Pope (Federal commander of the western armies) that he would not accept the same terms of surrender that had been offered to General Lee the month before.[66] After the Federal officer who had delivered the original proposal was briefed on Smith's decision and dismissed, the defiant general, then in command of over fifty thousand troops,[67] sent the following document to the governors of the western states:

Headquarters Trans-Mississippi Department, Shreveport, La., May 9, 1865. Their Excellencies Henry W. Allen, Governor of Louisiana; Pendleton Murrah, Governor of Texas; H. Flanagin, Governor of Arkansas and Thomas C. Reynolds, Governor of Missouri; Gentlemen: The surrender of General Lee, and the perilous situation of the armies in North Carolina and Alabama seem to preclude the probability of successful resistance in the states east of the Mississippi. The army under my command yet remains strong, fresh, and well equipped. The disparity of numbers, though great, between it and our enemies may be counterbalanced by valor and skill.

Under these circumstances it is my purpose to defend your soil and the civil and political rights of our people to the utmost extent of our resources, and to try and maintain untarnished the reputation which our soldiers have so nobly won on many fields. . . . The Trans-Mississippi Department is so separated from the states on the eastern side of the Mississippi that communication is suspended. Since the evacuation of Richmond, the seat of government for the Confederate States has not been fixed, and it may be transferred to the western side of the Mississippi. . . . I have therefore requested you to assemble in conference and ask you to indicate such policy as you may deem necessary to maintain with honor and success the sacred cause in which we are engaged. E. Kirby Smith, General.[68]

While General Smith outlined his reasons for continuing the war and requested that the western governors assemble to discuss the current situation, two Union regiments under the command of Col. Theodore H. Barrett landed near Brownsville, Texas, and marched toward the city. While advancing along the banks of the Rio Grande on the morning of May 12, Colonel Barrett's men surprised a Confederate outpost at Palmito Ranch and after several minutes of exchanging fierce fire forced them to abandon their position. Throughout the night of May 12 both sides brought up reinforcements, and by the early-morning hours of the following day, both Union and Confederate forces were evenly matched at about five hundred each.[69] At the commencement of the engagement on the next day, it became clear that the Confederates held the advantage, for they alone had a battery of artillery. Colonel Barrett made an official report of the engagement: "With the Rio Grande on our left, a superior force of the enemy in our front, and his flanking force on our right, our situation at this time was extremely critical. Having no artillery to oppose the enemy's six twelve-pounder pieces, our position became untenable. We therefore fell back fighting. This movement, always difficult, was doubly so at this time, having to be performed under a heavy fire from both front and flank.... Every attempt of the enemy's cavalry to break this line was repulsed with loss to him, and the entire regiment fell back with precision and perfect order."[70]

Years later, Col. John Salmon "Rip" (Rest in Peace) Ford (an ex-officer in the Texas Rangers who commanded the 2nd Texas Cavalry at the Battle of Palmito Ranch) appears to have had a different opinion of the engagement, for he wrote in his memoirs that the battle from the beginning had been "a run" and demonstrated well "how fast demoralized men could get over ground." Ford further remarked that Colonel Barrett "seemed to have lost his presence of mind" and led his troops off the field in a "rather confused manner."[71]

This last action of the American Civil War, unquestionably a Confederate victory, lasted about four hours and cost the Federals 111 officers and men captured and about 30 men killed or wounded (Confederate casualties were reported to have been little more than a handful). Colonel Barrett's official report of the engagement several months later to the adjutant general of the US Army includes his following observations about what he considered to be the last shots fired during the Civil War: "The last volley of the war, it is believed, was fired by the Sixty-second U.S. Colored Infantry about sunset on the 13th of May, 1865, between White's Ranch and the Boca Chica, Texas."[72]

Ironically, while the Battle of Palmito Ranch took place on the banks of the Rio Grande, the Confederate governors whom Gen. Kirby Smith had contacted some days earlier had reached a decision about continuing the war and informed their headstrong commander (contrary to his assurances that his army was capable of continuing the fight) that they wished him to enter into negotiations with the Federals.[73] General Smith had been given no choice and reluctantly sent emissaries to negotiate surrender terms.

At the initial meeting the Confederates learned that President Jefferson Davis had been taken into custody several days earlier in rural Georgia and that the Army of the Trans-Mississippi was then the only Confederate force still under arms. With a vast majority of government officials in irons, and an overwhelming veteran force poised to invade their territory as soon as logistics could be worked out, the officers whom Smith had sent to meet with the enemy returned to their commander with terms of surrender.[74] While news of Jefferson Davis's capture then circulated throughout the Trans-Mississippi and negotiations with Federal authorities were under way, Confederate officers found it increasingly difficult to keep the men under their command from simply going home.[75]

On May 26 Confederate military emissaries on behalf of Gen. E. Kirby Smith surrendered the Armies of the Trans-Mississippi to General E. R. S. Canby at New Orleans and immediately made plans for both General Smith and Magruder to officially surrender their commands in Galveston harbor several days later. On June 2, 1865, Confederate generals Smith and Magruder went aboard the USS *Fort Jackson* and put their signatures to the articles of surrender. Both generals immediately made plans to go south (along with about three or four thousand other Confederates who refused to live under Federal rule) and offer their services to Emperor Maximilian of Mexico.[76]

The story of the members of Captain Singer's Secret Service Company seems to have ended exactly where it began, along the same stretch of Buffalo Bayou where Edgar Singer and John Fretwell had first demonstrated their new contact mine to General Magruder some two and a half years earlier.[77] All founding members except Fretwell were then back in Texas, so the end must have seemed strange to Captain Singer's group, for unlike others in the defeated South they had continuously planned (virtually to the last hour of the last day of the conflict) to once again take the fight to the enemy and attempt to disperse its blockading fleet from within huge semi-submerged, iron-plated torpedo boats of their own invention.

On June 25, 1865 (forty-eight hours after the last Confederate general, Stand Watie, surrendered his Native American regiments to Federal forces), an inventory of Trans-Mississippi vessels that had been taken into Federal custody was compiled in Galveston. Only the torpedo boat that had been under construction at Lubbock's Mill appears on the list, and no word in this comprehensive report even hints that Union naval officers had ever seen the huge ironclad vessel known to have been built at Captain Chubb's shipyard.[78]

In the August 29, 1909, edition of the *Galveston Daily News*, former Singer Secret Service Corps member and local resident John D. Braman was interviewed about his involvement with Confederate torpedoes, the submarine *H. L. Hunley*, and building of the mysterious Galveston-based torpedo boats during the last year of the war:

> In October 1862, the Confederate Congress recognized the torpedo service, and a bureau was established at Richmond where a special corps of officers and men were raised.... One of the officers of this bureau, Major J. D. Braman, is still a resident of Galveston, and was connected with the filling out and designing of the first submarine....
>
> During this period there was a Confederate Navy Yard at Goose Creek, in the upper bay, in charge of the late Commodore Thomas H. Chubb. Here the construction of a torpedo boat was begun in the summer of 1864, by means of which it was proposed to destroy the federal blockading fleet at the entrance to the harbor. The work however, proceeded slowly, in consequence of the great scarcity of necessary material, and before the craft was completed the war had ended.[79]

If John Braman knew the fate of the ironclad torpedo boat constructed at Chubb's Galveston shipyard, he never revealed the information during the interview. Captain Singer's group may have secretly scuttled the vessel prior to Federal occupation. Nothing is known about the fate of the little Houston-based submarine boat described in M. P. Hunnicutt's March 1865 report, and the fact that the vessel may never have existed is a possibility.[80]

Federal forces were poised to occupy all regions of Texas in the weeks following General Magruder's surrender, and regimental commanders had ample time to either hide or destroy their various weaponry prior to disbanding. It is therefore reasonable to suppose that Capt. Edgar Singer and the men under his command may well have followed the example set by their fellow Texans and opted to destroy their own inventions rather than surrender them to the Federals.[81] Union general Phillip Sheridan filed a report regarding his opinions about how Texans had deceived the Federal government by disbanding before officially surrendering:

> On the 29th of May I assumed control of this new command, designating it the Military Division of the Southwest, with headquarters at New Orleans, La.; and at about the same time, received intelligence of the surrender of E. Kirby Smith, through commissioners sent from him to Major-General Canby. This surrender was made, but bore upon its face double dealing on the part of the rebel commander, or his agents, as the Texas troops had declined to surrender,

and had disbanded to their homes, destroying magazines and carrying with them arms and ammunition from the different arsenals.

General Smith proceeded to Galveston, and from thence escaped to Mexico, in violation of the agreement he had bound himself to observe. This conduct on his part may have arisen from the fact that it could not be concealed that his real object in offering to surrender was to get security for the Arkansas, Louisiana and Missouri troops to return to their homes, knowing full well that the Texas troops did not intend to surrender, and that most of them had already gone to their homes; that while they were destroying their arsenals and carrying home with them their arms, it was their constant boast that they were not conquered and that they would renew the fight at some future day.[82]

Texans never renewed "the fight at some future day," and most eventually signed parole documents at local provost marshal offices that Federal occupying forces established throughout the state. In early July 1865, Capt. Edgar Singer and the rest of the founding members of the Singer group walked into a recently established office at Port Lavaca and signed parole papers that stated they would not "hereafter serve in the armies of the Confederate States, or in any military capacity whatever, against the United States of America." With the signing of those federal documents on that hot July day, the remarkable story of the Singer Secret Service Corps (who had done more damage to the Union war effort than perhaps any ten Confederate regiments combined) officially came to an end.[83]

Conclusion

IN 1865, THE SECRETARY OF THE US NAVY, Gideon Welles, reported to Congress that the navy had lost more ships during the war from Confederate torpedoes than from all other causes combined.[1] The chief of the Confederate army's "Torpedo Bureau," Gen. Gabriel Rains, claimed after the war that the operatives under his command (including the men assigned to Captain Singer's group) damaged or sank some fifty-eight Federal ships during the war years.[2] Union records compiled soon after the end of the conflict state that about forty ships were either lost or damaged because of underwater mines.

Although the North made limited use of underwater mines, the Federals examined the torpedoes they captured with great interest, and most (after being disarmed) were forwarded to the Naval Ordnance Office for further study.[3] The West Point Museum in New York currently houses the largest collection of Confederate mines assembled during the war years. A brass plaque identifies one as a Singer torpedo dredged up from the Roanoke River in December 1864 (anchored there by John Fretwell and the Singer operatives under his command).

As the years passed, the horrors of the war slowly faded from the nation's collective memory. While the country turned its attention westward, the defeated veterans of the Confederate army affirmed their allegiance to the US government and went about rebuilding their lives and a new South. In the months following the surrender of Gen. Kirby Smith's Army of the Trans-Mississippi, the core members of Captain Singer's Secret Service Corps had returned to their homes in Port Lavaca. From 1870 census records, it appears that most of the original members lived no more than a few doors from one another. If the Singer group ever received bounties in gold for the numerous Federal ships they destroyed, it is not reflected in the 1870 census of Calhoun County, Texas. All the members mentioned in the census had limited means and, in some cases, seem to have been living in near poverty.

The Singer group may have often met socially to rehash their glory days and may have openly lamented the passing of B. A. (Gus) Whitney, who had died of pneumonia while working with the *Hunley* in Charleston Harbor. No mention of Whitney's wife and children appears in 1870 census documents.

Several months after the war had ended, John R. Fretwell, Edgar Singer's chief lieutenant, was able to make his way back to Texas. He retired from practicing medicine and opened a modest hotel in Port Lavaca. In 1867, Fretwell was elected Grand Master of Texas by his fellow Masons and by 1872 had sold his small Port Lavaca hotel and moved to Galveston. Several months later in 1873, Fretwell and his wife, Julia, moved to Mobile, a city he knew well from his service there during the war years. Fretwell's life in postwar Mobile appears to have been a hard one, for in 1883 at the annual convention of the Grand Lodge of Texas, the Masons adopted a resolution "to send John R. Fretwell, past Grand Master of this Grand Lodge, at once, the sum of three hundred dollars ($300), out of the grand charity fund, to be used in relieving him in his destitute condition."[4] Fretwell died in Mobile on April 17, 1885. The Masonic Grand Lodge of Texas paid for his fifty-dollar funeral.[5]

John D. Braman, part owner of the *Hunley* and member of the first Charleston crew who later oversaw construction of the ironclad torpedo boat at Lubbock's Mill, was paroled with Edgar Singer in Port Lavaca in early July 1865 and soon after opened a small grocery store. Some years later, Braman and his small family moved to Galveston, where in 1909 a reporter from the *Galveston Daily News* interviewed him about his wartime activities with Captain Singer's Secret Service Corps. Braman was not only a *Hunley* crew member (with the McClintock crew) but also helped contribute to the design of the submarine.[6] In 1912, John Braman died at his home in Galveston, and according to Edgar Singer, many wartime documents relating to Braman's involvement with the *Hunley* and the Singer organization were found among his private papers.[7]

James Jones, the talkative Singer operative who unknowingly informed the Union spy M. P. Hunnicutt about secretly built Singer submarines, appears to have given up his life as a jeweler after the war (what southerner could have afforded jewelry during Reconstruction?) and pursued carpentry instead.[8] No records regarding James Jones or his whereabouts are known after the 1870 census of Calhoun County, Texas.

C. E. Frary, Edgar Singer's brother-in-law and fellow member of the Port Lavaca Masonic Lodge who had accompanied Singer to the eastern regions of the Confederacy in early 1863, returned to Port Lavaca after the war and opened a combination repair shop and wagon-making facility.[9] Frary married Singer's sister prior to the war, so Edgar; his wife, Harriet; and their children, Richard, Amy, and George, most likely were frequent visitors to the Frary household during the turbulent years of Reconstruction.

Baxter Watson, co-designer of the *Pioneer* and submarine advocate who petitioned both Jefferson Davis and General Beauregard during the closing months of the war for funding to build an electrically powered submarine, never received the necessary money he requested for the electro-magnetic engine. Watson continued to work with other members of Captain Singer's group at the Park and Lyons machine shop in Mobile until the end of the war. Nothing is known of Baxter Watson's postwar activities. He died in New Orleans in his late thirties before 1872.[10]

Henry J. Leovy, the prewar friend and associate of Horace Hunley, who owned a share of both the privateering submarine *Pioneer* and the Singer-built *H. L. Hunley*, returned to practicing law in his hometown of New Orleans shortly after parting ways with the Confederate secretary of war, Judah Benjamin, in mid-May 1865. He eventually wrote several books on the Louisiana judicial system and represented ex-Confederate president Jefferson Davis in a legal action in 1879.[11] On October 9, 1902, Leovy collapsed on Canal Street from a sudden illness and died the following day. He was laid to rest at the Trinity Church in Pass Christian, Louisiana, next to his son, whom Leovy and his wife had named after his old friend Horace Lawson Hunley.[12]

B. J. Sage, one of the founders of the Confederate Secret Service whose captured letter of introduction to Singer operative Robert W. Dunn (which named various southern agents operating in occupied territories) almost compromised all covert operations along the Mississippi River, was working in Europe when the war ended. In the months that followed, he amassed funds and prepared a defense brief in anticipation of representing Jefferson Davis in his trial for treason. His brief, "The Republic of Republics" (a document that outlined in part the legalities surrounding a state's right to withdraw from the Union), was published in the United States around 1866. Sage died in reduced circumstances in Louisiana in 1902.[13]

Henry Dillingham, the ex-*Hunley* crewman sent by Jefferson Davis into occupied regions to destroy Federal supply depots and riverboats with his partners Edward Frazier and Thomas Clarke, was not heard from again after the group's dispersal near Memphis in late September 1864. By February 1865, most of the group had been captured, and the St. Louis provost marshal issued a warrant for Henry Dillingham's arrest for the burning of the medical supply depot in Louisville. In early June 1865, Dillingham's partner in sabotage, Edward Frazier, was taken to Washington in irons to testify at the Lincoln assassination trial regarding covert activities Jefferson Davis had ordered.

At the time of the trial Federal prosecutors hoped that high officials in the Confederate government could be implicated in the assassination conspiracy. By portraying Confederate covert operations in the worst light possible, they hoped to show Davis as a monster capable of ordering such an act. In an attempt to portray Henry Dillingham's past sabotage activities behind enemy lines in the darkest way imaginable, Federal prosecutors made the claim that Dillingham had not torched a medical supply depot in Louisville (as was recorded on his arrest warrant), but instead a Union hospital filled with wounded soldiers.[14] Frazier must have been stunned by this fabricated line of questioning, but with his own life in the balance (he had by then confessed to burning several boats with great loss of life), he appears to have gone along with the deception.

Since Henry Dillingham was a marked man, he perhaps took on a new identity and disappeared into postwar society with what remained of the gold he had received from the Confederate Treasury for torching the medical supply depot. Federal authorities never apprehended Dillingham, and we do not know what became of this Mobile-based steamboat engineer who had been a member of George Dixon's Charleston crew.[15]

James R. McClintock, who with his partners Henry Leovy, Baxter Watson, and Horace Hunley first envisioned building the *Pioneer* in New Orleans, returned to Mobile with his wife and children after the war and became the owner and captain of a local dredge boat.[16] In 1868, McClintock wrote to underwater explosives expert Matthew Maury (who was then on the faculty of the Virginia Military Institute) in an attempt to gain support for approaching the Europeans with his submarine idea. Within that letter McClintock gave a very interesting but brief history of the three submarines he helped construct.[17]

During the summer of 1872, McClintock attended secret meetings with representatives from the British navy in Nova Scotia, hoping to sell them his submarine diagrams. With letters of endorsement from both Matthew Maury and Gen. John Slaughter, the British were impressed but eventually passed on his offer.[18] In the autumn of 1879, while demonstrating to representatives of the US government a newly developed underwater contact mine of his own invention, something went amiss. The explosive device detonated prematurely and killed him instantly.[19]

Lt. William Alexander, the *Hunley*'s first officer while under the command of George Dixon, completed the breech-loading cannon for which he had been called back from Charleston to design before the end of the war. During the evacuation of Mobile in early April 1865, Alexander's revolutionary new cannon was put aboard a steamer that accompanied the last remnants of the Mobile garrison to the interior of Alabama. As the Union forces approached several days later, the gun was pushed overboard to avoid its capture.[20]

At the end of hostilities, Alexander returned to his prewar occupation as a consulting engineer and entered into a partnership with the owners of the Park and Lyons machine shop.[21] In the years that followed Alexander became a Thirty-Second Degree Mason, joined the local chapter of the United Confederate Veterans, and became the chief electrician for the city of Mobile. During the early years of the twentieth century he wrote two articles regarding his adventures in the *Hunley*, and thanks to his efforts, we now have what will probably prove to be the most complete history of the Confederate submarine ever to come to light written by someone who was actually there. William Alexander died at his residence at 13 South Catherine Street on May 13, 1914, some fifty years after being called away from his duties aboard the *Hunley*.[22]

The background of Lt. George E. Dixon, the detached engineering officer from the 21st Alabama who commanded the *Hunley* on the night of February 17, 1864, will perhaps remain a mystery. Records at the Probate Court in Mobile County, Alabama, show that on March 6, 1866 (over two years after Dixon's death aboard the *Hunley*), the court gave permission to the administrator of his estate to sell off Dixon's belongings at public auction. Accompanying the petition to auction Dixon's "Leather Trunk," then filled with what remained of his possessions, was a list of its contents. The articles within the trunk seem to have been somewhat desirable (especially during the lean years of Reconstruction), and in all likelihood he had no relatives to lay claim to his property. We still know little of Lt. George E. Dixon's prewar life.[23]

Robert W. Dunn, part owner of the *Hunley*, was ordered south from Richmond in early August 1863 to help with the operations of the group's submarine boat in Charleston. He later oversaw construction of huge ironclad torpedo boat at Captain Chubb's Galveston shipyard. Dunn was paroled in Port Lavaca in early July 1865 with fellow Masons Edgar Singer and John Braman. No mention of a Robert W. Dunn appears in either the 1870 or 1880 census records for Texas, so we know nothing about his postwar activities.[24]

Capt. Edgar Collins Singer, founder of the group of engineers who caused so much havoc with the Union war effort, returned to his home in Port Lavaca after General Magruder's surrender and continued to tinker with mechanical devices of every description. Information recorded in the 1870 census of Calhoun County reveals that Edgar Singer listed his postwar occupation as inventor and had a net worth of just over two hundred dollars. In 1874, Singer and his family moved to Marlin, Texas. In the years that followed the US government granted him various patents for "a number of appliances for cotton gins, and a brick mold," which in some regions of the South became the standard.[25]

During the summer of 1916, while World War One was being waged in Europe and the Imperial German Navy was practicing unrestricted submarine warfare, a reporter for the *San Antonio Express* heard rumors about the accomplishments of Captain Singer's group during the Civil War and contacted the ninety-one-year-old veteran to interview him. Horace Hill conducted the only known Singer interview to date, published as "Texan Gave World First Successful Submarine Torpedo" on July 30, 1916. Singer revealed for the first time information regarding the group's formation and involvement with the *Hunley*, and thanks to Hill's efforts, we now have a reasonably comprehensive history of Singer's secret organization.

Edgar Singer admitted that he and his fellow Port Lavaca Masonic Lodge members would never have gotten involved with underwater warfare had Federal gunboats not savagely shelled their hometown in late October 1862. If this event had never taken place, Singer and his friends probably would have spent the remainder of the war manning a relatively insignificant backwater garrison along the Texas coast. Singer indicated in his interview that he developed his unique underwater mine for the sole purpose of protecting Port Lavaca and his own family from future attack and never envisioned that one day his invention (designed and fabricated in his backyard machine shop) would become the most successful underwater contact mine in the Confederacy.[26] Shortly after the United States had entered World War One, Capt. Edgar Collins Singer, the founder of Singer's Secret Service Corps, died in Marlin at the age of ninety-three.[27]

Myth building in the postwar South had the effect of drawing attention away from the study of covert organizations that the Confederates had established during the Civil War.[28] Abraham Lincoln, in the years following his assassination, came to be viewed as one of the greatest Americans who ever lived, and to have suggested that high officials in the Confederate government may have been involved in plotting his death would have been devastating to the South's self-image of being a chivalrous people.[29] We can assume that it was in the South's best interest to block all avenues of study that might inadvertently link southern covert organizations to the barbaric actions of John Wilkes Booth.

For these reasons, we can also assume that ex-members of such secret organizations decided that it was in their collective best interest to remain anonymous after the war and not draw attention to the covert organizations to which they had once belonged. By keeping past group members and operations secret, no one could be accused of being involved in the infamous plot that took the life of Abraham Lincoln in early April 1865.[30] Thus, it seems obvious why no one has compiled a comprehensive study of Confederate covert organizations.[31]

Captain Singer may well have shared such views since he did not publicly reveal the many operations that his unit took part in during the war years until his 1916 interview.[32] Although the article revealed many previously unknown details regarding his secret organization, only his group's involvement with the *Hunley* submarine and the use of his patented contact mine were actually covered in any detail.

The topics of railroad torpedoes, huge ironclad torpedo boats, and the names of the various vessels sunk by Singer mines were noticeably absent from the San Antonio news article, so even at that late date, Singer may have had reservations about revealing everything he knew regarding the group's covert activities.[33] By 1866, officers of the US Corps of Engineers had determined that the Singer-Fretwell torpedo had done more damage to Federal warships than any other mine used by the South.[34]

From our own investigation we know that the Singer group was responsible for sinking the following vessels during the two years that they were in operation: USS *Baron DeKalb*, *Housatonic* (sloop-of-war sunk by the *Hunley* on February 17, 1864), *Eastport*, *Tecumseh*, *Narcissus* (Federal gunboat sunk at the mouth of the Dog River on December 8, 1864, and raised some days later), *Ostego*, *Bazely*, *Milwaukee*, and *Osage*. Of the nine known vessels sunk, five were ironclads. According to Federal records mines destroyed or damaged some thirty other vessels, and most likely several vessels not listed here were also victims of a Singer torpedo.

Judging from the number of Union vessels sunk by the Singer group, it seems that group members would have become rich beyond their wildest dreams, if in fact the Confederate government was paying the agreed 50 percent bounty on every vessel wrecked.[35] If the government was issuing these bounties, the Singer organization would have been entitled to more than one hundred thousand dollars for the sinking of the *Housatonic* alone, for the vessel's construction costs were known to have exceeded two hundred thousand dollars.[36] Because most of the group was living in reduced circumstances by 1870, it is fair to assume that these promised government bounties were either paid in Confederate scrip or were never forthcoming.[37]

During our investigation into the Singer Secret Service Corps, we may have inadvertently discovered evidence of a fascinating relationship between initiated group members and the fraternal order of Freemasons. The core members of Singer's unit were more than casually involved with the order. John Fretwell was known to have been a devoted member who shortly after the war was elected Grand Master of Texas. It seems reasonable to conclude that the affinity with freemasonry that existed between these men contributed to strengthening the bonds of Captain Singer's unique organization.[38] Capt. Horace Hunley and James McClintock were themselves initiated into the Masonic order in New Orleans prior to the South's secession, which may have contributed to their being inducted into Edgar Singer's secret group of engineers in early March 1863.

Lt. George Dixon's leather trunk of possessions remained unclaimed into the early spring of 1866. A quickly written inventory of the contents of the trunk that accompanied the legal documents in Dixon's estate file stated that the contents were civilian articles of clothing that had been left behind prior to his journey to Charleston.[39] Most of the items were rather ordinary and of little interest; however, at the bottom of the list appeared the scribbled notation "1 box Masonic books." Dixon's affiliation with the Freemasons was verified when an engraved Masonic emblem (excavated from within the hull of the *Hunley*) was found attached to his watch chain. The engraved inscription stated that Dixon had been an initiated member of Mobile's Lodge number 40. Unfortunately, many wartime Masonic records were lost soon after the conflict had ended, and thus far nothing regarding the background of George Dixon has come to light from Alabama's Masonic Archives.

In mid-July 1864, George Dixon's first officer, William Alexander, was himself inducted into Mobile's Masonic Lodge number 40, a strange turn of events considering that he was then struggling with the design of a new breech-loading cannon while a war was raging around him.[40] To learn the many secret teachings and oaths connected with the Freemasons during such a stressful, chaotic time seems only to reinforce the fact that membership in the Masonic order was highly regarded by those affiliated with the Singer organization. Although we can only speculate at this late date about how important these perceived bonds might have been to group members, it is interesting to note that the *Hunley*'s discoverer, Clive Cussler, was himself inducted into the Masonic order while serving in the military during the mid-1950s,[41] a unique piece of information that only adds to the mystique that already surrounds the fascinating history of the bold group of men who designed, fabricated, and manned the world's first successful combat submarine.

Notes

Chapter 1

1. Cotham, *Battle on the Bay*, 7.
2. Ibid., 1.
3. "Torpedo Inventor Dies," 1918 obituary of Edgar Collins Singer from the private collection of John Hunley. Copy filed in Hunley Archive, Warren Lasch Conservation Center.
4. Hill, "Texan Gave World First Submarine Torpedo."
5. Perry, *Infernal Machines*, 43.
6. Hill, "Texan Gave World First Submarine Torpedo."
7. Ibid.
8. *War of the Rebellion*, ser. 1, 4:112–13.
9. Ibid., 9:707.
10. Fitzhugh, "Saluria, Fort Esperanza," 76.
11. Cotham, *Battle on the Bay*, 52. Cotham cites this quote from Porter, "The Opening of the Lower Mississippi," 26.
12. *Official Records of the Union and Confederate Navies*, ser. 1, 18:60.
13. Eustace Williams Collection, Mobile Public Library; Henry J. Leovy "Biography," Henry J. Leovy Papers, Williams Research Center.
14. Henry J. Leovy Papers, Williams Research Center.
15. James McClintock letter to Matthew Fountain Maury, Matthew F. Maury Papers, vol. 46, item 9087–9094, Manuscript Division, Library of Congress.
16. Confederate States Records, Register of Commissions, Library of Congress; copy filed in Hunley Archives, Warren Lasch Conservation Center.
17. Ibid., 100.
18. Stern, *The Confederate Navy*, 101.
19. *Official Records of the Union and Confederate Navies*, ser. 1, 18:463.
20. *War of the Rebellion*, ser. 1, 9:710.
21. Fitzhugh, "Saluria, Fort Esperanza," 69, 73.
22. *War of the Rebellion*, ser. 1, 15:143–47.
23. *Official Records of the Union and Confederate Navies*, ser. 1, 19:213.
24. Cotham, *Battle on the Bay*, 65. Cotham cites as his source "Notice of four-day evacuation period dated October 4, 1862. Courtesy of Rosenburg Library, Galveston, Texas."
25. Ibid., 63–64.
26. Robert M. Franklin, "Battle of Galveston." Manuscript written by Franklin on April 2, 1911, for *United Confederate Veteran Magazine*, chapter 105. Available at the Library of Congress.
27. Ibid., 68.

28. Fitzhugh, "Saluria, Fort Esperanza," 80.
29. *War of the Rebellion*, ser. 1, 15:181–83.
30. Hill, "Texan Gave World First Submarine Torpedo."
31. Ibid.
32. 1860 Census, Calhoun County, Texas; and biography of Dr. John R. Fretwell, in "Transactions Texas (Masonic) Lodge of Research June 7, 1980–March 21, 1981, vol. 15." Copies in author's collection and Hunley Archive, Warren Lasch Conservation Center.
33. Biography of Dr. John R. Fretwell.
34. Fitzhugh, "Saluria, Fort Esperanza," 80.
35. Hill, "Texan Gave World First Submarine Torpedo."
36. Ibid.
37. Ibid.
38. Fitzhugh, "Saluria, Fort Esperanza," 81.
39. Cotham, *Battle on the Bay*, 90. Cotham cites Arnold, *Early Life and Letters of Thomas J. Jackson*, 95–96.
40. Fitzhugh, "Saluria, Fort Esperanza," 81.
41. Cotham, *Battle on the Bay*, 111.
42. Stern, *The Confederate Navy*, 124.
43. Hill, "Texan Gave World First Submarine Torpedo."
44. Ibid.
45. Ibid.
46. Order Book, Department of Texas, New Mexico, and Arizona, January–April 1863, Record Group 109, National Archives (hereafter NA).
47. Letters Received, Department of Texas, New Mexico, and Arizona, December 31, 1862–April 30, 1863, Record Group 109, Box 1, NA.
48. Von Scheliha, *Treatise on Coast-Defence*, 228.
49. Ibid.
50. Interviews with George Rhodes, historian of Calhoun County, Texas, 2001; and Jack Beeler, historian for the Port Lavaca Masonic Lodge, 2001.
51. Hill, "Texan Gave World First Submarine Torpedo."
52. Tidwell, *April 65*, 112.
53. Biography of Dr. John R. Fretwell, "Transactions Texas (Masonic) Lodge of Research June 7, 1980–March 21, 1981, vol. 15."
54. 1860 census, Calhoun County, Texas; and Hill, "Texan Gave World First Submarine Torpedo."
55. Interviews with George Rhodes and Jack Beeler; 1860 census, Calhoun County, Texas; and Hill, "Texan Gave World First Submarine Torpedo."
56. Hill, "Texan Gave World First Submarine Torpedo."
57. Bradbury's name appears on an early February 1863 invoice for $115 for Singer operative James Jones. Confederate Navy Subject File, Record Group 109, Reel 11, NA. Bradbury is also referred to as Captain Bradbury in several Department of Texas order books and is credited with being in charge of all Texas torpedo operations in Perry, *Infernal Machines*, 46.
58. Letters Received, Department of Texas, New Mexico, and Arizona, December 31, 1862–April 30, 1863, Record Group 109, Box 1, NA.
59. Letter from Major Daniel Shea, in ibid.

60. James Jones, $115.00 invoice, Confederate Navy Subject File, Record Group 109, M1091, Reel 11, NA.
61. Order Book, Department of Texas, New Mexico, and Arizona, January–April 1863, Record Group 109, NA.
62. William Longnecker's Confederate war record, Compiled Service Records of Confederate Soldiers Who Served in Organizations from the State of Texas, Record Group 109, "Shea's Texas Artillery," NA.
63. Perry, *Infernal Machines*, 46.
64. This would seem to have been the logical course of action. A document dated February 11, 1863, regarding the deployment of additional Singer men to General Smith's headquarters alludes to this: Special Order Number 70, Special Orders, Department of Texas, New Mexico, and Arizona, chap. 2, 3:113, NA.
65. Ibid.
66. Letter to Flag Officer Tucker from Admiral Buchanan, August 1, 1863, Buchanan Letter Books, Southern Historical Society Collection.
67. Hill, "Texan Gave World First Submarine Torpedo."
68. Special Order 31, Section 6, March 17, 1863, Special Orders East Sub-District of Texas, New Mexico and Arizona, Record Group 109, chap. 2, 102:109, NA.
69. Confederate Papers Relating to Citizens or Business Firms, M346, Reel 939, "E. C. Singer and Co.," Record Group 109, NA.
70. Hill, "Texan Gave World First Submarine Torpedo."
71. Tidwell, *April 65*, 83.
72. "Organization of Private Warfare: Bands of Destructionists and Captors," Confederate Navy Subject File, Record Group 109, NA.
73. Hill, "Texan Gave World First Submarine Torpedo."
74. "Torpedo Inventor Dies," 1918 obituary of Edgar Collins Singer from the private collection of John Hunley. Copy filed in Hunley Archive, Warren Lasch Conservation Center.
75. Confederate Papers Relating to Citizens or Business Firms, M346, "Leovy, Hunley and Duncan," October 12, 1862, Record Group 109, NA.
76. Letters Received by the Confederate Secretary of War, 1861–1865," Record Group 109, Entry M437, Reel 49, Image 872-G, NA.
77. Letter dated September 19, 1863, Confederate Papers Relating to Citizens or Business Firms, M346, "Horace L. Hunley" file, Record Group 109, NA.
78. Alexander, "True Stories of the Confederate Submarine Boats."
79. William Alexander speech to Iberville Historical Society, December 15, 1903, Mobile City Museum.
80. Letter from James McClintock to Matthew Fountain Maury, Maury Papers, vol. 46, item 9087–9094, Manuscript Division, Library of Congress.
81. Letter sent from Baxter Watson to Jefferson Davis, October 10, 1864, Letters Received by the Confederate Secretary of War, 1861–1865, Record Group 109, Entry M437, NA.
82. Letter from James McClintock to Matthew Fountain Maury, Maury Papers, vol. 46, item 9087–9094, Manuscript Division, Library of Congress.
83. Letter to the Confederate Secretary of the Navy, February 14, 1863, Buchanan Letter Books, Southern Historical Society Collection.
84. "Treasury of Early Submarines," 102.
85. Confederate deserter's testimony, Navy Area File, Record Group 45, Entry M625, Area 8, "January 1864," NA.

86. Letter from James McClintock to Matthew Fountain Maury, Maury Papers, vol. 46, item 9087–9094, Manuscript Division, Library of Congress.
87. Alexander, "True Stories of the Confederate Submarine Boats."
88. Hill, "Texan Gave World First Submarine Torpedo."
89. "Biographical Information on Henry J. Leovy, Close Associate of Horace Lawson Hunley and Part Owner of Submarine Torpedo Boat H. L. Hunley," Eustace Williams Collection at the Mobile Public Library.
90. Leovy may have taken over H. L. Hunley's share of the *Hunley* submarine after Hunley's death. Documents dealing with this appear with Horace Hunley's will, St. Tammany Parish Court House, Louisiana; letter from Leovy to General Beauregard, George E. Dixon's Confederate war record, Compiled Service Records of Confederate Soldiers Who Served in Units from the State of Alabama, Record Group 109, NA.
91. Hill, "Texan Gave World First Submarine Torpedo." Additional information regarding this partnership is filed in the Eustace Williams Collection, Mobile Public Library.
92. Alexander, "True Stories of the Confederate Submarine Boats." The *Hunley* submarine has features that Alexander does not mention. Alexander may have described a vessel that incorporated features that were unique to the *American Diver*.
93. Hill, "Texan Gave World First Submarine Torpedo."
94. Evidence of Dixon having been assigned to the *Hunley* is found in Dixon's Confederate war record, Compiled Service Records of Confederate Soldiers Who Served in Organizations from the State of Alabama, Record Group 109, 21st Alabama, NA. Additional information regarding this temporary assignment comes from a letter written by Dixon's company commander on August 8, 1863, cited in Folmar, *From That Terrible Field*, 117.
95. Confederate Papers Relating to Citizens or Business Firms, "John R. Fretwell" file, Record Group 109, NA.
96. Brown, "Confederate Torpedoes in the Yazoo," 3:580.
97. Biography of Dr. John R. Fretwell, "Transactions Texas (Masonic) Lodge of Research June 7, 1980–March 21, 1981, vol. 15."
98. A Hunley letter written in early May from Mississippi appears with documents filed in Horace Hunley's will now on file at the St. Tammany Parish Court House, Louisiana (copies in author's collection and Hunley Archive, Warren Lasch Conservation Center). Additional evidence placing Hunley in Mississippi comes from a Henry Leovy letter in Letters Received by the Confederate Secretary of War, 1861–1865, Record Group 109, Entry M437, NA. Leovy states in the letter that "Captain Hunley is now in Mississippi on the staff of General Maury."
99. A Hunley letter written in early May from Mississippi appears with documents filed in Horace Hunley's will, St. Tammany Parish Court House, Louisiana (copies in author's collection and Hunley Archive, Warren Lasch Conservation Center).
100. Confederate Papers Relating to Citizens or Business Firms, "E. C. Singer and Co.," Record Group 109, NA.
101. Letter from Robert W. Dunn to General Magruder, April 6, 1864, Confederate Papers Relating to Citizens or Business Firms, "R. W. Dunn" file, Record Group 109, NA.
102. Hill, "Texan Gave World First Submarine Torpedo"; Letters Sent by Confederate Engineering Department, Record Group 109, NA.
103. Letters Sent by Confederate Engineering Department, Record Group 109, NA. Although these order books appear complete, they contain nothing about meetings or mission assignments, so Singer must have been receiving his orders from a yet-undetermined source.

104. Tidwell, *April 65*, 83.
105. B. J. Sage, "Organization of Private Warfare," Confederate Navy Subject File, Record Group 109, Reel 11, NA. Copy on file in Hunley Archive, Warren Lasch Conservation Center.
106. Captured document appearing with General Order Number 85, US Mississippi Squadron, March 21, 1864, in *Official Records of the Union and Confederate Navies*, ser. 1, 26:191–92.
107. Tidwell, *April 65*, 104.
108. Each instance where the individual named is addressed as "captain" will be cited in future. Documents signed by Capt. Horace Hunley in Unfiled Papers and Slips Belonging in Confederate Compiled Service Records, Record Group 109, NA.
109. Special Order Number 259, Paragraph 10, Special Orders Issued by the Confederate Adjutant and Inspector General's Office, Record Group 109, NA.
110. Hill, "Texan Gave World First Submarine Torpedo."
111. Captain Courtenay's authorization document to raise a "Secret Service Corps" filed in Confederate Navy Subject File, Entry M1091, Reel 12, Record Group 109, NA. This extremely rare document is a blank induction form meant to be signed by all individuals who wish to join "Captain Courtenay."
112. Tidwell, *April 65*, 48.
113. Ibid., 26.
114. The secretary of war promoted the commander of McDaniel's Secret Service Corps, Zedekiah McDaniel, to captain "without pay": "Z. McDaniel is hereby authorized to enlist a company of men, not to exceed fifty in number, for secret service against the enemy, under the regulations prescribed by this Department for such organizations. When he shall have enlisted and mustered his company into the service for the war he will receive a commission of Captain in the Provincial Army of the Confederate States, without pay. Transportation will be furnished to him for his recruits to the place of rendezvous and to such points as he may select for his operations." *War of the Rebellion*, ser. 4, 3:177. From evidence in this document, "Capt." Edgar Singer received a similar promotion "without pay" from the secretary of war.
115. The first known document bearing the name "Singer's Secret Service Corps" appears in a report dated March 10, 1864, in Letters Received by the Confederate Adjutant and Inspector General's Office, Record Group 109, NA.
116. Captain Courtenay's authorization document dated August 18, 1863, to raise a "Secret Service Corps," in Confederate Navy Subject File, Entry M1091, Reel 12, Record Group 109, NA.
117. Report filed by Col. James Duff, January 11, 1864, with enclosed report from Capt. David Bradbury regarding past operations on the Texas coast, Confederate Papers Relating to Citizens or Business Firms, "D. Bradbury" file, Record Group 109, NA.
118. Ibid.
119. Von Ehrenkrook, *History of Submarine Mining and Torpedoes*, 34.
120. Report filed by Col. James Duff, January 11, 1864, with enclosed report from Capt. David Bradbury regarding past operations on the Texas coast, Confederate Papers Relating to Citizens or Business Firms, "D. Bradbury" file, Record Group 109, NA.
121. Letter sent by the Confederate commissioner of patents to E. C. Singer, May 18, 1863, Letters Sent by the Confederate Commissioner of Patents 1863, Museum of the Confederacy.
122. Confederate Papers Relating to Citizens or Business Firms, "E. C. Singer and Co.," Record Group 109, NA.

123. Letter to General Magruder, April 6, 1864, Confederate Papers Relating to Citizens or Business Firms, "R. W. Dunn" file, Record Group 109, NA.
124. Ibid.
125. Ibid.
126. R. W. Dunn invoice for $1,350 drawn at Shreveport, Louisiana, May 29, 1863, Confederate Navy Subject File, Record Group 109, Reel 11, NA.

Chapter 2

1. Letter from General Leadbetter, July 1, 1863, in Letters and Telegrams Sent, Engineer Office at Mobile, June–July 1863, Record Group 109, chap. 3, vol. 10, NA.
2. Invoice for twenty Singer torpedoes, June 30, 1863, Confederate Papers Relating to Civilian or Business Firms, Whitney and Braman file, Record Group 109, Entry M346, NA; letter from General Leadbetter, July 1, 1863, Letters and Telegrams Sent, Engineer Office at Mobile, June–July 1863, Record Group 109, chap. 3, vol. 10, NA.
3. Maury, "How the Confederacy Changed Naval Warfare," 80; Alexander, "Heroes of the *Hunley.*"
4. Flato, *The Civil War*, 113.
5. Ibid.
6. Filed with Hunley's will in St. Tammany Parish, Louisiana; copies filed in Hunley Archive, Warren Lasch Conservation Center.
7. *Official Records of the Union and Confederate Navies*, ser. 1, 25:281.
8. Ibid.
9. Ibid., 286.
10. Brown, "Confederate Torpedoes in the Yazoo," 3:580.
11. *Official Records of the Union and Confederate Navies*, ser. 1, 25:283.
12. Ibid.
13. Ibid.
14. Ibid., 286.
15. Confederate Papers Relating to Citizens or Business Firms, "John R. Fretwell" file, Record Group 109, NA.
16. Fitzhugh, "Saluria, Fort Esperanza," 88.
17. Edgar Singer letter, February 8, 1864, Confederate Navy Subject File, Record Group 109, Reel 11, NA; J. D. Braman letter, Letters Received by the Secretary of the Navy from Commanding Officers of Squadrons, Record Group 45, Entry M89, Reel 132, Image 380, NA. For an abbreviated copy of this letter, see *Official Records of the Union and Confederate Navies*, ser. 1, 26:187.
18. Confederate Navy Subject File, Record Group 109, Reel 11, NA. This document also discusses how the Singer device could be used in railroad sabotage.
19. Letter dated July 22, 1863, Letters Sent by the Confederate Engineering Department, Record Group 109, Entry M628, NA.
20. Dunn order sending him to Charleston from Richmond on August 10, 1863, in Confederate Navy Subject File, Record Group 109, Reel 11, NA; document dated August 22, 1863, in "C. E. Frary" file, Compiled Service Records of Soldiers Who Served in Organizations from the State of Texas, Record Group 109, NA. James Jones stayed in Richmond, according to a short note in Letters Sent by the Confederate Engineering Department, Record Group 109, Entry M628, NA: "September 22, 1863, Mr. James Jones, the bearer of this, is an employee of the Engineer Department on duty in this city and vicinity."

21. Within the Hunley File, Mobile Historical Preservation Society, is an index card listing several *Hunley*-related sources of information not in its collection. One note mentions that the signature of Horace L. Hunley is found in the ledger book of the Battle House Hotel for September 12, 1863 (location of ledger unspecified). Photocopy of this index card filed in Hunley Archive, Warren Lasch Conservation Center.
22. Invoice for eighty Singer torpedoes, July 25, 1863, in Confederate Papers Relating to Civilian or Business Firms, Whitney and Braman file, Record Group 109, Entry M346, NA. Gen. James Slaughter wrote a letter stating that the *Hunley* was being tested prior to July 31, 1863. General J. E. Slaughter File, Compiled Service Records of Confederate General and Staff Officers, and Non-Regimental Enlisted Men, Record Group 109, NA. Additional information regarding early tests of the *Hunley* also found in a postwar Slaughter letter, in file ADM 1/6236/39455, "Submarine Warfare," British Admiralty Archives; copies of all documents filed in Hunley Archive, Warren Lasch Conservation Center.
23. Postwar Slaughter letter, in file ADM 1/6236/39455, "Submarine Warfare," British Admiralty Archives.
24. Ibid.
25. Letter to Flag Officer Tucker, August 1, 1863, Franklin Buchanan Letter Books, Southern Historical Society Collection.
26. Hill, "Texan Gave World First Submarine Torpedo." Within this article Singer stated that J. D. Braman was placed in charge of the *Hunley* and accompanied it to Charleston. Other contemporary sources indicate that Whitney was actually in charge.
27. Gen. J. E. Slaughter File, Compiled Service Records of Confederate General and Staff Officers, and Non-Regimental Enlisted Men, Record Group 109, NA.
28. Letter to Flag Officer Tucker, in Franklin Buchanan Letter Books, Southern Historical Society Collection.
29. Navy Area File, Record Group 45, Entry M625, NA.
30. Telegrams Sent, Department of South Carolina, Georgia, and Florida, 1863–1864, Record Group 109, NA.
31. Ibid.
32. Information revealed in testimony given by Confederate deserters Shipp and Belton, Navy Area File, Record Group 45, Entry M-625, Area 8, NA.
33. Folmar, *From That Terrible Field*, 117, 118. This book contains other information regarding George Dixon.
34. Letter dated August 8, 1863, Ellen Shackelford Gift Papers, Southern Historical Society Collection.
35. Nichols, *Confederate Engineers*, 68.
36. Order filed in Letters Sent, Department of South Carolina, Georgia, and Florida, 1863–1864, Record Group 109, NA.
37. Confederate Navy Subject File, Record Group 109, Reel 11, NA; Braman's order in Letters Sent by Confederate Engineering Department, Record Group 109, NA.
38. Letter dated July 22, 1863, Letters Sent by Confederate Engineering Department, Record Group 109, Entry M628, NA.
39. Document dated August 22, 1863, "C. E. Frary" file, Compiled Service Records of Soldiers Who Served in Organizations from the State of Texas, Record Group 109, NA.
40. Johnson, "Jeremiah Donovan."
41. Letters Sent, Department of South Carolina, Georgia, and Florida, 1863–1864, Record Group 109, NA.

42. Augustine Smyth Communications from 1890s, South Carolina Historical Society.
43. Middleton Papers, South Carolina Historical Society.
44. Letters Sent by the Commissioner of Patents, Museum of the Confederacy; photocopy filed in Hunley Archive, Warren Lasch Conservation Center.
45. Information obtained from an August 5, 1863, document directed to the Confederate secretary of war: "Something similar has been brought forward recently by two different persons, McDaniel and Singer, both place the torpedo under the track which operating by percussion blows up the first train that passes." Confederate Navy Subject File, Record Group 109, Reel 11, NA; photocopy filed in Hunley Archive, Warren Lasch Conservation Center.
46. Horace Hunley folder, Unfiled Papers and Slips Belonging in Confederate Compiled Service Records, Record Group 109, NA.
47. News clipping found among the Augustine Smyth Communications from 1890s, South Carolina Historical Society.
48. Miscellaneous Papers, Department of South Carolina, Georgia, and Florida, "Telegrams Received," June–October 1863, Box 49, NA.
49. Middleton Papers, South Carolina Historical Society.
50. Kloeppel, *Danger beneath the Waves*, 35.
51. Stanton, "Submarines and Torpedo Boats," 398–99.
52. *Official Records of the Union and Confederate Navies*," ser. 1, vol. 28, pt. 2, 670.
53. Ibid.
54. Documented information presented throughout this chapter proves that all the named individuals were present in Charleston by mid-August. Since the *Hunley* could be operated with a crew of eight, the named Singer operatives may well have made up the crew. It is also likely that various unnamed individuals may also have been rotated as crew members throughout the operation. All the named men must have had some experience within the vessel, for they all had invested a substantial amount of money into the venture.
55. *Official Records of the Union and Confederate Navies*, ser. 1, vol. 28, pt. 2, 670.
56. Order dated August 26, 1863, directing new personnel (Private C. L. Sprague) to be attached to the submarine boat, in Letters Sent, Department of South Carolina, Georgia, and Florida, 1863–1864, Record Group 109, NA.
57. Stanton, "Submarines and Torpedo Boats," 398–99.
58. Letters Sent, Department of South Carolina, Georgia, and Florida, 1863–1864, Record Group 109, NA.
59. Hill, "Texan Gave World First Submarine Torpedo." The 1860 census of Calhoun County, Texas, lists B. A. Whitney as having five children, three boys and two girls.
60. Wilkinson, "The Peripatetic Coffin."
61. Telegrams Sent, Department of South Carolina, Georgia, and Florida, 1863–1864, Record Group 109, NA.
62. Middleton Papers, South Carolina Historical Society.
63. Stanton, "Submarines and Torpedo Boats," 398–99.
64. Metzger, "Brother Charles Hazelwood Hasker and the 'Fish,'" Collection of Dr. Charles Perry; copy filed in Hunley Archive, Warren Lasch Conservation Center.
65. Middleton Papers, South Carolina Historical Society.
66. *Charleston Daily Courier*, August 30, 1863. A recently discovered payroll book of the CSS *Palmetto State* for the months of July and August 1863 at the National Archives has the following entry penned next to the name of Absolum Williams: "Drowned in submarine battery on 29th August, 1863." Typed copy in Eustace Williams Collection,

Special Collections Division, Mobile Public Library. Original (photocopy) of payroll file for CSS *Palmetto State* (Record Group 109, NA) also filed in Hunley Archive, Warren Lasch Conservation Center.

67. Union report on Charleston torpedoes, March 7, 1865, a federal document compiled after the close of the war, in Confederate Navy Subject File, Record Group 109, Reel 11, NA; photocopy filed in Hunley Archive, Warren Lasch Conservation Center.
68. Union debriefing of Capt. M. M. Gray, 1865, Confederate Navy Subject File, Record Group 109, Reel 11, NA.
69. "Torpedo Warfare," by General Rains, Museum of the Confederacy.
70. Order dated September 3, 1863, P. G. T. Beauregard Private Papers, Record Group 109, Entry 116, NA.
71. Telegrams Sent, Department of South Carolina, Georgia, and Florida, 1863–64, Record Group 109, NA.
72. Confederate Papers Relating to Civilian or Business Firms, Smith and Broadfoot file, Record Group 109, Entry M346, NA.
73. Ibid.
74. Within the Hunley File in the Mobile Historical Preservation Society is an index card listing several *Hunley*-related sources of information. One note mentions that the signature of Horace L. Hunley is found in the ledger book of the Battle House Hotel for September 12, 1863 (location of ledger unspecified). Photocopy of this index card filed in Hunley Archive, Warren Lasch Conservation Center.
75. Letter to General Beauregard, September 19, 1863, Confederate Papers Relating to Civilian or Business Firms, Horace L. Hunley file, Record Group 109, Entry M346, NA.
76. Ripley to Jordan, Letters Received, Department of South Carolina, Georgia, and Florida, 1863–1864, Record Group 109, NA.
77. Navy Area File, Record Group 45, Entry M625, Area 8, Confederate States Navy, Reel 414, NA; photocopy filed in Hunley Archive, Warren Lasch Conservation Center.
78. "Alphabetical List of Patentees for the Year 1863," Museum of the Confederacy.
79. Letter from E. C. Singer, February 28, 1864, Confederate Navy Subject File, Record Group 109, Reel 11, NA.
80. Confederate Papers Relating to Citizens or Business Firms, "Horace L. Hunley" file, RG 109, Entry M-346, NA.
81. Beauregard, "Torpedo Service," 5, 4.
82. Letters Sent, Department of South Carolina, Georgia, and Florida, 1863–64, Record Group 109, NA.
83. Alexander, "True Stories of the Confederate Submarine Boats."
84. Henry Dillingham and a man named Marshall are mentioned as crew members in an 1899 letter written by William Alexander, in Augustine Smyth Communications from 1890s, South Carolina Historical Society.
85. "Dixon: Builder of the Submarine Hunley, Went to Death in the Deep."
86. Dixon's Confederate war record, 21st Alabama Regiment, Compiled Service Records of Soldiers Who Served in Organizations from the State of Alabama, Record Group 109, NA.

Chapter 3

1. "The city is filled with great men, there has been as high as 6 or 8 generals stopping at the Spotswood House where we are at all times." Edgar Singer letter, February 8, 1864, Confederate Navy Subject File, Record Group 109, Reel 11, NA.

2. Pinkerton, *Spy of the Rebellion*, 414–28.
3. Ibid.
4. Summary of Jones letter, Letters Received by the Navy Department, Record Group 45, Entry M-517, Box 23, Number 312, NA: "In the same mail I find a letter dated Richmond, Virginia. October 13th 1863, written by James Jones, who seems to be a practical machinist and informs his correspondent that the week previous he 'set ten torpedoes in the Pauneky River and have fourteen more ready which will be set this week. I have 74 very nearly done sir, I have my orders for Wilmington and start tonight at 4 ½ o'clock.'"
5. Still, *Confederate Navy*, 135.
6. Beauregard, "Torpedo Service," 5, 4.
7. Stanton, "Submarines and Torpedo Boats," 398–99.
8. *Official Records of the Union and Confederate Navies*, ser. 1, 15:692.
9. Confederate Papers Relating to Citizens or Business Firms, "Angus Smith" file, Record Group 109, Entry M-346, NA.
10. Orders and Circulars, Department of South Carolina, Georgia, and Florida, September 1863–March 1864, Record Group 109, NA.
11. Letters Sent, Department of South Carolina, Georgia, and Florida, Record Group 109, NA.
12. *Official Records of the Union and Confederate Navies*, ser. 1, 15:693.
13. Flato, *The Civil War*, 42–43.
14. *War of the Rebellion*, ser. 1, vol. 31, pt. 3, 10.
15. Ibid., 220.
16. Document directed to Confederate secretary of war, August 5, 1863: "Something similar has been brought forward recently by two different persons, McDaniel and Singer, both place the torpedo under the track which operating by percussion blows up the first train that passes." Confederate Navy Subject File, Record Group 109, Reel 11, NA; photocopy filed in Hunley Archive, Warren Lasch Conservation Center.
17. Confederate Papers Relating to Civilian or Business Firms, Robert W. Dunn file, Entry M-346, Record Group 109, NA. This was written in mid-April 1864, some months after the destruction of the trains and corroborated in Federal documents in *War of the Rebellion*, vol. 31, *Operations in Kentucky, Southwest Virginia, Tennessee, Mississippi, North Alabama, and North Georgia, October 20–December 31, 1863*. Several acts of railroad sabotage utilizing contact torpedoes identical to those of the Singer design did take place in Tennessee during the months of October–December 1863. Although the exact number of locomotives blown from Tennessee tracks could not be determined from the federal documents, Dunn's claim of eight is well within possible limits.
18. *War of the Rebellion*, ser. 1, vol. 31, pt. 1, 740.
19. Ibid., 755.
20. Ibid., pt. 3, 587.
21. Ibid., 16.
22. Ibid.
23. Ibid., vol. 38, pt. 4, 579.
24. Crowley, "The Confederate Torpedo Service," 290.
25. The first attack seems to have taken place on October 26. By December patrols of the rail line were established, and reports regarding acts of sabotage against the line gradually disappear from the official records.
26. *War of the Rebellion*, ser. 1, vol. 31, pt. 3, 474.

27. Ibid., 473.
28. "Scottsborough, January 12, 1864. Brig. General J. E. Smith: A torpedo was exploded under the track on the railroad today, without, however, doing any damage, save the cross-ties." Ibid., vol. 32, pt. 2, 74.
29. Confederate Papers Relating to Civilian or Business Firms, Robert W. Dunn file, Entry M-346, Record Group 109, NA.
30. "After considerable hesitation, we were finally ordered by the Secretary of War, to construct one boat at Selma, Alabama, and one at Wilmington, N.C., of the following dimensions vis. 160 feet long, 28 foot beam and 11 foot hold with flat deck, carrying all their machinery below—to be iron sheathed and with no capacity for guns, and only showing 2 feet above water when ready for work. They [are] to be arranged with torpedoes, worked from below decks, and through tubes, forward, Aft, and on both sides. It is believed by Engineers of the highest rank, after a full investigation of our plans, that these boats will be perfectly able to raise the Blockade of all the Harbors in the Confederacy." Ibid.
31. Letter from Gardner Smith to Mrs. V. W. Barrow, Hunley File, Mobile City Museum.
32. Beauregard, "Torpedo Service," 5, 4.
33. Alexander, "True Stories of the Confederate Submarine Boats."
34. Letters Sent, Department of South Carolina, Georgia, and Florida, Record Group 109, NA.
35. Orders and Circulars, Department of South Carolina, Georgia, and Florida, September 1863–March 1864, Record Group 109, NA.
36. Letter from Gardner Smith to Mrs. V. W. Barrow, Hunley File, Mobile City Museum.
37. "Last Honors to a Devoted Patriot."
38. Perry, *Infernal Machines*, 46.
39. Fitzhugh, "Saluria, Fort Esperanza," 94.
40. Ibid., 95
41. Ibid.
42. Ibid., 97
43. King, *Torpedoes*, 5, 6. David Bradbury reported that "in relation to the Singer torpedo: In the month of May last, and by order of Major Shea, I placed 18 floating torpedoes in the channel between Fort Esperanza and the bar at Pass Cavallo." Confederate Papers Relating to Civilian or Business Firms, David Bradbury file, Entry M-346, Record Group 109, NA.
44. Confederate Papers Relating to Civilian or Business Firms, David Bradbury file, Entry M-346, Record Group 109, NA.
45. King, *Torpedoes*, 5, 6.
46. Fitzhugh, "Saluria, Fort Esperanza," 97.
47. Confederate Papers Relating to Civilian or Business Firms, David Bradbury file, Entry M-346, Record Group 109, NA.
48. Ibid.
49. Telegrams Sent and Orders, Department of South Carolina, Georgia, and Florida, Record Group 109, chap. 2, vol. 45, NA.
50. George E. Dixon Estate File, Probate Court, Mobile County, Alabama.
51. "Dixon, Builder of the Submarine Hunley, Went to Death in the Deep"; Hartwell, "An Alabama Hero."
52. Lt. George E. Dixon, Company "A," 21st Regiment Alabama Volunteers, Compiled Service Records of Soldiers Who Served in Organizations from the State of Alabama,

Record Group 109, NA. Since the story of the bullet-stopping gold coin originated in a Mobile newspaper, the author of the article may have obtained the fanciful tale from someone who had been close to Dixon and was well aware of the facts, perhaps William Alexander, who is known to have lived in Mobile at the time.
53. Discovered by *Hunley* archeologist Maria Jacobson.
54. Endorsements on Letters Received, Department of South Carolina, Georgia, and Florida, 1863–64, Record Group 109, NA.
55. Ibid.
56. Letters Sent, Department of South Carolina, Georgia, and Florida, 1863–64, Record Group 109, NA.
57. Telegrams Sent, Department of South Carolina, Georgia, and Florida, 1863–64, Record Group 109, NA.
58. Letters Received, Department of South Carolina, Georgia, and Florida, 1863–64, Record Group 109, NA.
59. "November, 13, 1863. From E. C. Singer and Company. Report of Operations of Torpedo Company," Index of Letters Received by the Confederate Engineering Department, Record Group 109, NA.
60. Report in Confederate Papers Relating to Civilian or Business Firms, Robert W. Dunn file, Entry M-346, Record Group 109, NA: "We were at once transferred by him to the Engineer Troops and ordered to report to General Joseph E. Johnston at Morton, Mississippi. Leaving operators at Richmond, Wilmington, Savannah, Mobile and Charleston." Although no other source of Singer-related information places men in Savannah, Dunn seems sure that this was the case.
61. Summary of a captured Jones letter, Letters Received by the Navy Department, Record Group 45, Entry M-517, Box 23, Number 312, NA.
62. "We at once concentrated all the inventive genius in our party for the purpose of getting up something new that would carry destruction to the Yankees, make money for ourselves, and at the same time be of great service to the Confederacy. The result was that I got up the plan of an ironclad torpedo boat that, all who saw it admitted, was equal to the task of destroying any war ship now afloat." Letter from J. D. Braman to his wife, March 3, 1864, Confederate Navy Subject File, Reel 11, Record Group 109, NA.
63. Cotham, *Battle on the Bay*, 185. Evidence that Robert Dunn and Captain Singer met with Senator Wigfall in letter from J. D. Braman to his wife, March 3, 1864, Confederate Navy Subject File, Reel 11, Record Group 109, NA: "We deemed it best, after consultation, to send Dunn to Richmond, and through the influence of Wigfall and others to get the matter [the construction of ironclad torpedo boats] before Congress."
64. Two known Confederate documents designate Singer's torpedo organization as being assigned to the Confederate Secret Service. The first states that James McClintock and Baxter Watson were both members: Report, March 10, 1864, in Letters received by the Confederate Adjutant and Inspector General's Office, Entry N-1224-1864, Record Group 109, NA. The second is a late-war communication regarding the redeployment of James Jones to Texas: "Head Quarters 3rd Corps. A.N.V. Eng. Dept. Petersburg 11th Nov. 1864. Brig. Gen. W. H. Stevens Chief Eng. General: . . . He [James Jones] had been put by the Eng. Bureau on the same footing with Singer, Dunn & Braman—but—had no written notice to that effect though told by Mr. Dunn that he had been assigned to Captain Singer's Secret Service Company." Letters Received by the Confederate Secretary of War, Entry 1475-J-1864, Record Group 109, NA.

65. "While operating in the later place (Charleston) it was fully demonstrated, that Torpedoes could be used on the prow, placed on the bow of a boat, without any damage to herself, yet carrying certain destruction to the vessel attacked. On the first of November last, the writer returned again to Richmond, to obtain assistance for the construction of such boats as would enable us to operate with safety to our crews, and at the same time, strike terror to the enemy." Letter to General Magruder, April 6, 1864, in Confederate Papers Relating to Civilian or Business Firms, Robert W. Dunn file, Entry M-346, Record Group 109, NA.
66. Letter from J. D. Braman to his wife, March 3, 1864, Confederate Navy Subject File, Reel 11, Record Group 109, NA.
67. Letter from Dunn to Engineering Department, December 23, 1863, Letters Sent by the Confederate Engineering Department, 1861–65, Entry M-628, Record Group 109, NA.
68. Edgar Singer letter dated February 8, 1864, Confederate Navy Subject File, Record Group 109, Reel 11, NA.
69. Alexander, "Heroes of the *Hunley*."
70. Beauregard to Ingraham, December 11, 1863, Letters Sent, Department of South Carolina, Georgia, and Florida, 1863–64, Record Group 109, NA.
71. Confederate Navy, the First Submarine Boats, filed in "North Carolina Leaf," 16–18, Museum of the Confederacy; photocopy filed in Hunley Archive, Warren Lasch Conservation Center.
72. Lieutenant Dixon had three volunteers with him upon his return: "January 10th 1864. The Chief Quartermaster will cause Lt. G. E. Dixon to be refunded his actual expenses for lodging and subsisting for four men in Charleston attached to the Submarine Torpedo Boat or Engine of War, from the 12th of November to the 16th of December 1863, being the sum of six hundred and thirty one (631) dollars." Orders and Circulars, Department of South Carolina, Georgia, and Florida, September 1863–March 1864, Record Group 109, NA.
73. General Orders, Department of South Carolina, Georgia and Florida, July 1862–January 1864, Record Group 109, NA.
74. Alexander, "Heroes of the *Hunley*."
75. Letter from James Tomb, June 7, 1908, Collection of Dr. Charles Perry; photocopy in Hunley Archive, Warren Lasch Conservation Center.
76. Alexander, "True Stories of the Confederate Submarine Boats."
77. Letters Sent by the Confederate Engineering Department, 1861–65, Entry M-628, Record Group 109, NA.
78. Ibid.
79. April 6, 1864, letter to General Magruder, Confederate Papers Relating to Civilian or Business Firms, Robert W. Dunn file, Entry M-346, Record Group 109, NA; letter from J. D. Braman to his wife, March 3, 1864, Confederate Navy Subject File, Reel 11, Record Group 109, NA.
80. Summary of captured Jones letter, Letters Received by the Navy Department, Record Group 45, Entry M-517, Box 23, Number 312, NA.
81. Letters Sent by the Confederate Engineering Department, 1861–65, Entry M-628, Record Group 109, NA.
82. From the previously quoted December 9, 1863, letter to Louisiana congressman Henry Marshall, it could be argued that James McClintock may never have left Charleston after the government seized the submarine. No documentation indicates that he ever left the city: "James Island, S.C. December 9th, 1863. My Friend: In

compliance with my promise, I went yesterday to the city and hunted up McClintock the owner I believe of the submarine boat. I interrogated him in reference to the fatal mishap that attended Hunley in his last experimental trip." Furman, Green, and Chandler papers, Manuscript Department, Library of Congress.

83. "Head Quarters 3rd Corps. A.N.V. Eng. Dept. Petersburg 11th Nov. 1864. Brig. Gen. W. H. Stevens Chief Eng. General: I have the honor to submit for your action the following—James Jones—a private in Shea's Battalion Heavy Artillery Texas Coast was detached some eighteen months ago for duty with E. C. Singer & Co.—in the manufacture and use of the Singer Torpedo. With others he was sent by Maj. Gen. Magruder to Lt. Gen. E. K. Smith—and by him this side of the Mississippi River. He has been in the employ of the Engineer Department in this vicinity and Wilmington, but under the recent order I presume he must be ordered back to his Company." Letters Received by the Confederate Secretary of War, Entry 1475-J-1864, Record Group 109, NA.

84. Letter from J. D. Braman to his wife, March 3, 1864, Confederate Navy Subject File, Reel 11, Record Group 109, NA.

85. Bradford, *History of Torpedo Warfare*, 46; Perry, *Infernal Machines*, 199. Both these sources list by date and location all Union vessels sunk or damaged by Confederate torpedoes during the Civil War, and neither reports any actions as having ever taken place on the York River, Virginia. Both Bradford and Perry got their information from Federal naval records prior to publication of their books.

86. Letter to General Magruder, April 6, 1864, Confederate Papers Relating to Civilian or Business Firms, Robert W. Dunn file, Entry M-346, Record Group 109, NA.

87. Telegrams Sent, Department of South Carolina, Georgia, and Florida, 1863–64, Record Group 109, NA.

88. Navy Area File, January 1, 1864–January 31, 1864, Record Group 45, Entry M-625, Area 8, NA.

89. Ibid.

90. *Official Records of the Union and Confederate Navies*, ser. 1, 15:334.

91. Alexander, "The True Stories of the Confederate Submarine Boats."

92. Letters Sent, Department of South Carolina, Georgia, and Florida, 1863–64, Record Group 109, NA.

93. Alexander, "True Stories of the Confederate Submarine Boats."

94. Confederate Papers Relating to Civilian or Business Firms, David Bradbury file, Entry M-346, Record Group 109, NA.

95. Letters received by the Secretary of the Navy from Commanding Officers of Squadrons, January–March, 1864, RG 45, Entry M-89, NA.

96. Letter written by James Tomb, June 7, 1908, Collection of Dr. Charles Perry; photocopy in author's collection and Hunley Archive, Warren Lasch Conservation Center.

97. "Remarkable Career of a Remarkable Craft."

98. Letter from Francis Lee to General Beauregard, May 15, 1876, Eustace Williams Collection, Miscellaneous Papers, Special Collections Division, Mobile Public Library. On the upper corner of this letter appears "Original letter in possession of Dr. Thomas Macmillan, Philadelphia, Pa." Photocopy in Hunley Archive, Warren Lasch Conservation Center.

99. Hill, "Texas Gave World First Submarine Torpedo," provides additional proof regarding Beauregard's involvement with reconfiguring the *Hunley*'s torpedo: "Experiments at Charleston brought out the fact that there was not enough water under the keels of the outlying union vessels to permit the submarine to pass below them; whereupon

General Beauregard changed the arrangement of the torpedo by fastening it to the bow."

100. Von Kolnitz, "The Confederate Submarine." From Captain Hunley's previously quoted rope requisition of October 10, 1863, the actual length was most likely 150 feet, not 150 yards.
101. Alexander, "True Stories of the Confederate Submarine Boats."
102. Orders and Circulars, Department of South Carolina, Georgia, and Florida, September 1863–March 1864, Record Group 109, NA.
103. Alexander "True Stories of the Confederate Submarine Boats."
104. Letters Received by the Secretary of Navy from Commanding Officer of Squadrons, 1841–86, Record Group 45, Entry M89, NA.
105. Alexander, "True Stories of the Confederate Submarine Boats."
106. Ibid.
107. Ibid.
108. Ibid.
109. Ibid.
110. Alexander, "Heroes of the *Hunley*," 749.
111. Telegrams Sent and Orders, Department of South Carolina, Georgia, and Florida, Record Group 109, chap. 2, vol. 45, NA.
112. Alexander, "True Stories of the Confederate Submarine Boats."
113. Alexander, "Heroes of the *Hunley*," 749.
114. Alexander letter in Augustine Smyth Communications from 1890s, South Carolina Historical Society; photocopy in Hunley Archive, Warren Lasch Conservation Center.
115. Orders and Circulars, Department of South Carolina, Georgia, and Florida, Record Group 109, NA.
116. Hill, "Texan Gave World First Submarine Torpedo."
117. "Richmond. February 8, 1864. Dear Wife: I wrote you yesterday a short letter by a Mr. Adams that lives at Harrisburg, six miles below Houston, Mr. Dunn leaves for Texas tomorrow." Edgar Singer letter dated February 8, 1864, in Confederate Navy Subject File, Record Group 109, Reel 11, NA. We know that Dunn stopped over in Mobile before attempting to cross the Mississippi River: "Mobile, Al., March 3, 1864. My Dear Wife: I write this to send by Bob Dunn, who leaves here Saturday morning for home." Ibid. Additional proof comes from Dunn himself on April 6, 1864: "In Feb. last the writer was ordered to report to Lt. Gen. E. K. Smith of the Trans Miss. Dept., for torpedo service, with instructions to him to furnish us with such assistance as he might think the service required in his Dept. In crossing the Mississippi River on the night of the 16th instant, in company with Col. Ward and Col. Clark, I had the misfortune to lose my papers, having been closely pursued by launches from a gunboat, and fired at three times from a small swivel gun on their bows, before nearing the shore and twice afterwards." Confederate Papers Relating to Civilian or Business Firms, Robert W. Dunn file, Entry M-346, Record Group 109, NA.
118. Confederate Navy Subject File, Record Group 109, Reel 11, NA. Additional copy with added paragraphs in Letters Received by the Secretary of the Navy from Squadron Commanders, Record Group 45, Entry 89, Reel 132, Image 394, NA.
119. Cardozo, *Reminiscences Charleston*, 124–25.
120. Proceedings of the Naval Court of Inquiry, February 26, 1864, Naval Records Collection, Record Group 45, Entry M-625, Area 8, Case Number 4345, NA.
121. Alexander, "True Stories of the Confederate Submarine Boats."

122. "I have a note from Dixon dated the day he went out for the last time." Alexander letter filed with Augustine Smyth Communications from 1890s, South Carolina Historical Society.
123. Proceedings of the Naval Court of Inquiry, February, 26, 1864, Naval Records Collection, Record Group 45, Entry M-625, Area 8, Case Number 4345, NA.
124. Cardozo, *Reminiscences Charleston*, 124–25.
125. Proceedings of the Naval Court of Inquiry, February, 26, 1864, Naval Records Collection, Record Group 45, Entry M-625, Area 8, Case Number 4345, NA.
126. Alexander, "True Stories of the Confederate Submarine Boats."
127. Proceedings of the Naval Court of Inquiry, February, 26, 1864, Naval Records Collection, Record Group 45, Entry M-625, Area 8, Case Number 4345, NA.
128. Alexander, "True Stories of the Confederate Submarine Boats."
129. Proceedings of the Naval Court of Inquiry, February, 26, 1864, Naval Records Collection, Record Group 45, Entry M-625, Area 8, Case Number 4345, NA.
130. Ibid.
131. Ibid.
132. Ibid.
133. Ibid.
134. Ibid.
135. Ibid.
136. *Official Records of the Union and Confederate Navies*, ser. 1, 15:332.
137. Proceedings of the Naval Court of Inquiry, February, 26, 1864, Naval Records Collection, Record Group 45, Entry M-625, Area 8, Case Number 4345, NA.
138. James Tomb discussed the spar assembly in a letter dated June 7, 1908: "Lieutenant Dixon and myself discussed the best plan to make use of the torpedo boat and decided that the adjustable spar used on the David was the best, and he said he would use it on the Hunley, keeping her on the surface and the torpedo some eight feet below." Collection of Dr. Charles Perry; photocopy in Hunley Archive, Warren Lasch Conservation Center. William Alexander provides additional proof in 1902 that the spar was angled several feet below the surface: "Did McClintock or any one see the torpedo boat after the explosion going out to sea apparently uninjured? No! If the Hunley has not yet been found, she can be found buried under the Housatonic. Or. And this is the question for (in my opinion) the Government to determine by experiment. Would an explosion of 100 pounds of powder attached to, and 25 feet from a submarine torpedo boat, ten feet below the surface injure the boat? If it would, and the Hunley is not under the Housatonic, then she is somewhere in the neighborhood of it." Letter Received (Hunley "Z" File), Naval Archives, Washington Navy Yard; photocopy in Hunley Archive, Warren Lasch Conservation Center.
139. Proceedings of the Naval Court of Inquiry, February, 26, 1864, Naval Records Collection, Record Group 45, Entry M-625, Area 8, Case Number 4345, NA.
140. *Official Records of the Union and Confederate Navies*, ser. 1, 15:332.

Chapter 4

1. It seems reasonable to assume that a signal fire was utilized because it could be seen from any point offshore. A signaling device that gave off a beam would be useless if the *Hunley* traveled north or south.
2. *Official Records of the Union and Confederate Navies*, ser. 1, 15:327.
3. Ibid., 328.

4. Ibid., 327.
5. Ibid., 330.
6. Ibid., 331.
7. Letters Received, Department of South Carolina, Georgia, and Florida, 1862–64, Record Group 109, NA.
8. Telegrams Sent, Department of South Carolina, Georgia, and Florida, 1863–64, Record Group 109, NA.
9. *Charleston Daily Courier*, February 29, 1864.
10. Telegrams Sent, Department of South Carolina, Georgia, and Florida, 1863–64, Record Group 109, NA.
11. Letter dated March 4, 1864, Middleton Papers, South Carolina Historical Society.
12. There is additional proof of these rumors :"February 28th, 1864. My dear little sister: There is a good piece of news here. The submarine torpedo boat was not lost, but had to go into Georgetown on account of the head wind, and she is there safe. She sunk the 'Housatonic,' a splendid sloop of war, built since the war. Is this not fine?" Smyth Papers, South Carolina Historical Society. John Clongh, a Confederate deserter, had heard the same rumors: "Name: John Clongh, Born: Ireland. Came to South Carolina in 1850, in Charleston five years. Worked on breastworks in forts of the rear, never worked on Sumter. Did not know in Charleston that the 'Housatonic' was sunk until boats crew of 'Nipsic' reported it. He has no knowledge of arrival back of torpedo boat. Report was that it went in at Georgetown." Testimony of John Clongh, Area File of Naval Records Collection, Record Group 45, Entry M625, Area 8, NA.
13. Alexander, "True Stories of the Confederate Submarine Boats."
14. Proceedings of the Naval Court of Inquiry, Area File of Naval Records Collection, Record Group 45, Entry M625, Area 8, Case Number 4345, NA.
15. Letter from Henry Leovy filed in Lieutenant Dixon's war record, 21st Alabama, Compiled Service Records of Confederate Soldiers Who Served in Organizations from the State of Alabama, Record Group 109, Entry M-311, NA. Henry Leovy was to receive five thousand dollars from Hunley's estate upon his death. Because Horace Hunley owned a five thousand–dollar share of the submarine, Leovy may have turned down the cash and instead chosen to take over Captain Hunley's share of the boat. Documents dealing with this issue filed with Horace Hunley's will, St. Tammany Parish Court House, Louisiana; photocopies of all documents in Hunley's will filed in Hunley Archive, Warren Lasch Conservation Center.
16. Letter from Henry Leovy filed in Lieutenant Dixon's war record, 21st Alabama, Compiled Service Records of Confederate Soldiers Who Served in Organizations from the State of Alabama, Record Group 109, Entry M-311, NA.
17. *Official Records of the Union and Confederate Navies*, ser. 1, 15:337.
18. *War of the Rebellion*, ser. 1, 26:501.
19. B. J. Sage issued a sensitive document to Dunn, Singer, and Braman listing some fifty Confederate operatives working along the Mississippi River and throughout Louisiana, for reasons deduced from the opening and closing lines of the document: "To introduce R. W. Dunn, E. C. Singer and J. D. Braman to my friends . . . We must all help one another, and those who can be efficient in our cause must receive all necessary hospitality, aid, and information. I introduce none but the worthy. B. J. Sage." Letters Received by the Secretary of the Navy from Squadron Commanders, Record Group 45, Entry M89, Reel 132, Image 406, NA; copy appears in *Official Records of the Union and Confederate Navies*, ser. 1, 26:192.
20. *War of the Rebellion*, ser. 1, 26:501.

21. Sage to Boggs, Confederate Papers Relating to Civilian or Business Firms, B. J. Sage file, Entry M-346, Record Group 109, NA; copy filed in Hunley Archive, Warren Lasch Conservation Center.
22. Ibid.
23. The only documentation regarding Colonels Ward and Clark is found in Unfiled Papers and Slips Belonging in Confederate Compiled Service Records, Record Group 109, NA. The documents state that both men commanded Missouri "Guerrilla Bands" on the western shore of the Mississippi River. Photocopies in Hunley Archive, Warren Lasch Conservation Center.
24. Dunn to Magruder, Confederate Papers Relating to Civilian or Business Firms, R. W. Dunn file, Entry M-346, Record Group 109, NA; copy in Hunley Archive, Warren Lasch Conservation Center.
25. Dunn Clark and Ward left for Texas on March 8: "Mobile, Alabama. March 7th 1864. My Dear Father: I expect that I will have an opportunity tomorrow to send a letter over the river. Colonel Ward, an old friend of mine expects to start and will go direct to Houston, as he will be careful with my letter I write with more hope that you will receive it." Letters Received by the Secretary of the Navy from Squadron Commanders, Record Group 45, Entry M89, Reel 132, NA.
26. Ibid.; abbreviated copy appears in *Official Records of the Union and Confederate Navies*, ser. 1, 26:187.
27. Letters Received by the Secretary of the Navy from Squadron Commanders, Record Group 45, Entry M89, Reel 132, NA; copy appears in *Official Records of the Union and Confederate Navies*, ser. 1, 26:184.
28. Document dated March 10, 1864, in Confederate Papers Relating to Civilian or Business Firms, E. C. Singer file, Entry M-346, Reel 939, Record Group 109, NA; photocopy in Hunley Archive, Warren Lasch Conservation Center.
29. Dunn to Magruder, Confederate Papers Relating to Civilian or Business Firms, R. W. Dunn file, Entry M-346, Record Group 109, NA; photocopy in Hunley Archive, Warren Lasch Conservation Center.
30. Hill, "Texan Gave World First Submarine Torpedo," documents that Captain Singer was acquainted with T. E. Courtenay :"The appointment of additional corps, commission and compensation being similar to those granted the Singer band [were established]. Prominent among these inventors were T. E. Courtenay, General Rains, Colonel Hill, Major Perkins, Hunter, Weldon and others."
31. Letters Received by the Secretary of the Navy from Squadron Commanders, Record Group 45, Entry M89, Reel 132, Image 396, NA; copy appears in *Official Records of the Union and Confederate Navies*, ser. 1, 26:186; Thatcher, "The Courtenay Coal Torpedo."
32. All the captured documents appearing in the group sent to the secretary of the navy were attributed to Dunn, Clark, and Ward. The only document that alludes to Colonel Ward is Dunn's letter to his father dated March 7, 1864 (quoted previously). The letter given to Colonel Ward included militarily significant information and therefore was included with the recovered Dunn and Clark documents. Letters Received by the Secretary of the Navy from Squadron Commanders, Record Group 45, Entry M89, Reel 132, NA; and Hunley Archive, Warren Lasch Conservation Center.
33. Letters Received by the Secretary of the Navy from Squadron Commanders, Record Group 45, Entry M89, Reel 132, NA; copy appears in *Official Records of the Union and Confederate Navies*, ser. 1, 26:185.
34. Ibid., 192.
35. Ibid., 184.

36. Ibid.
37. Ibid., 192.
38. Tidwell, *April 65*, 104.
39. Confederate Papers Relating to Civilian or Business Firms, Robert W. Dunn file, Entry M-346, Record Group 109, NA; photocopy filed in Hunley Archive, Warren Lasch Conservation Center.
40. Perry, *Infernal Machines*, 47.
41. Selfridge, "Retreat with Honor."
42. Letters Sent, Department of the Trans-Mississippi, February–August 1864, Record Group 109, NA.
43. Special Order 113, Special Orders, Department of Texas, New Mexico, and Arizona, April–June 1864, Record Group 109, NA.
44. Testimony of Union spy M. P. Hunnicutt. A description of these submarines appears in *Official Records of the Union and Confederate Navies*, ser. 1, 22:103–4. An additional, more comprehensive description of these vessels is filed in Office of the Chief Signal Officer, District of the Gulf and Western Mississippi, March 1865, Box 4, Number 1407, NA.
45. *Official Records of the Union and Confederate Navies*, ser. 1, 22:103–4.
46. Tidwell, *April 65*, 50–51.
47. Singer to Seddon, Letters Received by the Confederate Secretary of War, 1861–65, Record Group 109, Entry M-437, Number 755-T-64, NA.
48. Letters Received by the Confederate Adjutant and Inspector General's Office, 1861–65, Record Group 109, NA; photocopy filed in Hunley Archive, Warren Lasch Conservation Center.
49. Fretwell seems to have been in charge of the Richmond operation during this period: "Johnston's Headquarters, May 21, 1864, Dr. J. R. Fretwell, Care of Colonel W. H. Stevens, Sir: Can you raise some men and bring down the eight torpedoes via Drewry's Bluff? I can detail a few men here. I want them assigned for floating down upon the monitors that are shelling us. McDaniel has disappointed me. John A. Williams, Lieutenant-Colonel Engineers." *War of the Rebellion*, ser. 1, 36:818. If Singer was still at Richmond, it seems reasonable to assume that he received this order, not Fretwell. Singer may have returned to Mobile at about this time: "June 24, 1864, To E. C. Singer, Care of Lieutenant-Colonel Von Scheliha, Chief Engineer Mobile, Alabama. Sir: Details for yourself, Braman and others forwarded today by mail. A. L. Rives, Colonel." Letters Sent by the Confederate Engineering Department, Record Group 109, Entry M628, NA.
50. Flato, *The Civil War*, 148.
51. On July 25, 1864, Gen. Robert E. Lee stated in a handwritten letter to Gen. R. S. Ewell that "Dr. Fretwell was brought to me by Lieutenant-Colonel Williams, of the engineers, who stated that Dr. Fretwell was acting under the chief engineer at Richmond. He applied for and received my permission to operate on James River so far as that permission was necessary." Lee stated in another communication of the same date: "Doctor Fretwell is acting under no special orders from me. He was directed to report to General Rains." These documents seem to reinforce the theory that Captain Singer had indeed returned to Mobile, for Fretwell appears to have taken charge of the Virginia operation. *War of the Rebellion*, ser. 1, vol. 40, pt. 1, 796; and ibid., pt. 3, 800.
52. Ibid., vol. 36, pt. 1, 988.
53. Ibid., 745; Tidwell, *April 65*, 48.
54. Bradford, *History of Torpedo Warfare*, 8.

55. *Official Records of the Union and Confederate Navies*, ser. 1, 10:54.
56. Ibid.
57. *War of the Rebellion*, ser. 1, vol. 36, pt. 2, 745.
58. Crowley, "The Confederate Torpedo Service."
59. *War of the Rebellion*, ser. 1, vol. 40, pt. 3, 800.
60. "He served during the war on General Lee's staff, and was also connected with the setting of torpedoes in Mobile Bay." Fretwell obituary, *Mobile Register*, April 19, 1885.
61. *War of the Rebellion*, ser. 1, vol. 40, pt. 3, 800.
62. Ibid., vol. 36, pt. 3, 818.
63. Tidwell, *April 65*, 48. Tidwell footnotes this quote: "Rives to Colonel Stevens, May 15, 1864, Files of Charles T. Mason, Manuscript Collection, Virginia Historical Society."
64. Flato, *The Civil War*, 148–52.
65. Ibid.
66. "Mound City, June 25, 1864. Lieutenant-Commander F. M. Ramsey, Commanding USS Choctaw and 3rd District Mississippi Squadron. Sir: The rebels are fitting out at Shreveport four torpedo boats. They will be ready in two months" (*Official Records of the Union and Confederate Navies*, ser. 1, 26:438). Debriefing of Union spy Hunnicutt: "The boat is 40 feet long.... On the sides are two iron flanges (called fins) for the purpose of raising and lowering the boat in the water.... The boat is usually worked seven feet underwater and has four deadlights for the purpose of steering or taking observations.... The vessel is already at Houston and there are four others at Shreveport" (Office of the Chief Signal Officer, District of the Gulf and Western Mississippi, March 1865, Box 4, Number 1407, NA).
67. Special Orders Department of the Trans-Mississippi, 1864, Received by the District of Texas, Record Group 109, chap. 2, vol. 78, NA.
68. *Official Records of the Union and Confederate Navies*, ser. 1, 22:124; Testimony of a Confederate deserter, January 2, 1865, Area File of Naval Records Collection, Record Group 45, Entry M625, NA; photocopy of testimony filed in Hunley Archive, Warren Lasch Conservation Center.
69. Confederate Papers Relating to Civilian or Business Firms, Robert W. Dunn file, Entry M-346, Record Group 109, NA; photocopy filed in Hunley Archive, Warren Lasch Conservation Center.
70. Von Scheliha, *Treatise on Coast-Defence*, 317.
71. From a page copied from the Texas Order Books at the National Archives comes a list of military personnel engaged at various locations throughout Texas. Next to two names on a list of military personnel is the following notation about where the men are on duty: "Rocket Factory in Galveston." Special Orders, Department of Texas, New Mexico, and Arizona, 1864, Record Group 109, NA; photocopy filed in Hunley Archive, Warren Lasch Conservation Center.
72. Special Orders, Department of Texas, New Mexico, and Arizona, April–June 1864, Record Group 109, NA.
73. *Official Records of the Union and Confederate Navies*, ser. 1, 2:104; *War of the Rebellion*, ser. 1, vol. 48, pt. 1, 1197.
74. Letters Received, Department of Texas, New Mexico, and Arizona, January 1864–May 1865, Record Group 109, chap. 2, vol. 119, NA.
75. Special Orders, Department of Texas, New Mexico, and Arizona, July–September 1864, Special Order 190, Record Group 109, NA.
76. The title of "Captain Dunn" appears in a report filed by Gen. J. G. W. Walker, De-

cember 2, 1864, Letters Sent, Department of Texas, New Mexico, and Louisiana, September 1864–May 1865, Record Group 109, NA; photocopy filed in Hunley Archive, Warren Lasch Conservation Center.

77. Special Orders, Department of Texas, New Mexico, and Arizona, April–June 1864, Record Group 109, NA.
78. Ibid.
79. *War of the Rebellion*, ser. 1, vol. 26, pt. 2, 7.
80. Special Orders, Department of Texas, New Mexico, and Arizona, April–June 1864, Record Group 109, NA.
81. Confederate Papers Relating to Civilian or Business Firms, D. Bradbury file, Entry M-346, Record Group 109, NA; photocopy in Hunley Archive, Warren Lasch Conservation Center.
82. Report filed by Gen. J. G. W. Walker, dated December 2, 1864, Letters Sent, Department of Texas, New Mexico, and Arizona, September 1864–May 1865, Record Group 109, NA.
83. July 2, 1864, Book of Endorsements, District of Texas, New Mexico, and Arizona, April–August 1864, Record Group 109, NA.
84. Special Orders, Department of Texas, New Mexico, and Arizona, April–June 1864, Record Group 109, NA.
85. Cotham, *Battle on the Bay*, 68.
86. Special Orders Department of Texas, New Mexico and Arizona, July–September 1864, Record Group 109, National Archives.
87. *Official Records of the Union and Confederate Navies*, ser. 1, 26:438.
88. Letters Sent by the Confederate Engineering Department, Record Group 109, Entry M-628, NA.
89. General Walker to Chubb, February 21, 1865, Letters Sent, Department of Texas, New Mexico, and Arizona, Record Group 109, NA.
90. Perry, *Infernal Machines*, 115–17.
91. *War of the Rebellion*, ser. 1, vol. 40, pt. 1, 795–96.
92. Ibid., pt. 3, 800.
93. Ibid., pt. 1, 796.
94. Flato, *The Civil War*, 148–52.
95. With George McClellan as the Democratic Party's presidential candidate, its platform for the 1864 election openly called for an armistice with the Confederacy. Federal armies were bogged down in Virginia and northern Georgia during the spring of 1864, the public had grown weary of the war, and many historians think that Lincoln would have lost his bid for a second term if the election had been held several months earlier. Several Union victories during the late summer and early fall of 1864 improved northern morale just enough to reelect Lincoln to a second term. The Lincoln administration's policy was that there would be no negotiations with the Confederacy and the conflict would not be concluded until the South lay down arms, accepted the Emancipation Proclamation, and returned to the Union.
96. Tidwell, *April 65*, 133–35.
97. J. Andrews to Inspector General's Office, Letters Received by the Confederate Adjutant and Inspector General's Office, 1861–65, Entry N-1224-1864, Record Group 109, NA. The document lists all members of the Singer Secret Service Corps then on duty in Mobile. Both James McClintock and Baxter Watson's names head the list of four Singer operatives then on duty (nothing is known of the other two individuals named,

Greenleaf Andrews and John A. King, who were probably newcomers to the group). In an Engineering Department document penned on the same day as the Battle of Mobile Bay, James McClintock is referred to as being "in charge of torpedoes." Colonel Von Scheliha to McClintock, Letters Sent by Engineering Department, District of the Gulf, Record Group 109, NA; photocopies filed in Hunley Archive, Warren Lasch Conservation Center.

98. Von Scheliha, *Treatise on Coast-Defence*, 229.
99. Bradford, *History of Torpedo Warfare*, 53.
100. King, *Torpedoes*, 13.
101. "Torpedo Warfare," by General Rains, Museum of the Confederacy.
102. Perry, *Infernal Machines*, 159.
103. Von Ehrenkrook, *History of Submarine Mining and Torpedoes*, 46.
104. Maury, *Brief Sketch of the Work of Matthew Fontaine Maury*, 27.
105. Stern, *The Confederate Navy*, 55.
106. The location of the *American Diver* has not yet been discovered.
107. Stern, *The Confederate Navy*, 205.
108. Ibid., 204–11.
109. Schafer, *Confederate Underwater Warfare*, 148; and Scharf, *History of the Confederate States Navy*, 560.
110. Scharf, *History of the Confederate States Navy*, 561.
111. Maury, "How the Confederacy Changed Naval Warfare," 78.
112. Stern, *The Confederate Navy*, 205.
113. Perry, *Infernal Machines*, 161. Perry points out that some scholars believe that Farragut never said these words. For in-depth coverage of this debate, see Perry's citations.
114. Maury, "How the Confederacy Changed Naval Warfare," 78.
115. Perry, *Infernal Machines*, 161.
116. Bradford, *History of Torpedo Warfare*, 71; and Von Ehrenkrook, *History of Submarine Mining and Torpedoes*, 47.
117. Stern, *The Confederate Navy*, 209–10.
118. Letters Sent by Engineering Department, District of the Gulf, Record Group 109, NA; photocopy filed in Hunley Archive, Warren Lasch Conservation Center.
119. *Official Records of the Union and Confederate Navies*, ser. 1, 22:559.
120. Von Ehrenkrook, *History of Submarine Mining and Torpedoes*, 45; and Maury, *Brief Sketch of the Work of Matthew Fontaine Maury*, 28.
121. *Official Records of the Union and Confederate Navies*, ser. 1, 21:613.

Chapter 5

1. "Organization of Private Warfare," Confederate Navy Subject File, Record Group 109, NA.
2. Letter from General Maury to the Adjutant and Inspector General discussing activities of the Courtenay and McDaniel organizations, August 28, 1864, Letters Received by the Confederate Adjutant and Inspector General's Office, 1861–65, Record Group 109, Entry M474, Number 3972-B-64, NA.
3. *War of the Rebellion*, ser. 1, vol. 22, pt. 2, 697.
4. Records of the Office of the Judge Advocate General (Army), Correspondence, Letters Received, 1854–94, Record Group 153, Miscellaneous Correspondence, Box 12, Entry 6, NA. Information regarding Joseph W. Tucker's affiliation with the Confeder-

ate Secret Service obtained from enclosed February 22, 1865, Testimony of Confessed Boat Burner William Murphy, "Baker Report on Boat Burners."
5. Rule, *Joseph W. Tucker and the Boat Burners*, 3. Rule cites as the source of this information "W. Tucker to Jefferson Davis. From Confederate Memorial Hall, Spotswood Hotel 14th March, 1864. Confidential statements; for the President alone."
6. Ibid.
7. Records of the Office of the Judge Advocate General (Army), Record Group 153, Miscellaneous Correspondence, Box 12, Entry 6, NA. Information regarding the torching of river boats obtained from enclosed February 22, 1865, Testimony of Confessed Boat Burner William Murphy, "Baker Report on Boat Burners."
8. "Mr. Dillingham had been hired by General Polk and sent to Louisville, expressly to do that work (burn federal supply depots)." Investigation and Trial Papers Relating to the Assassination of President Lincoln, Record Group M599, Reel 13, page 3815 (subpage 52), "Testimony of Edward Frazier," NA.
9. Order dated October 16, 1863, Orders and Circulars, Department of South Carolina, Georgia, and Florida," Record Group 109, NA.
10. The 1860 census of Mobile, Alabama, lists Henry Dillingham's occupation as "Engineer." A description of Henry Dillingham as a "Steamboat Engineer" with residence at "Mobile, Alabama," enclosed with "Baker Report on Boat Burners," Records of the Office of the Judge Advocate General (Army), Record Group 153, Miscellaneous Correspondence, Box 12, Entry 6, NA.
11. Testimony of confessed Boat Burner William Murphy, February 22, 1865, in ibid.
12. Porter, "Torpedo Warfare,"
13. Thatcher, "The Courtenay Coal Torpedo," 5.
14. Bradford, *History of Torpedo Warfare*, 62.
15. Testimony of confessed Boat Burner George Goldthwaite, "Baker Report on Boat Burners," Records of the Office of the Judge Advocate General (Army), Record Group 153, Miscellaneous Correspondence, Box 12, Entry 6, NA.
16. Testimony of confessed Boat Burner William Murphy, in ibid.
17. Investigation and Trial Papers Relating to the Assassination of President Lincoln, Record Group M599, Reel 13, page 3815 (subpages 50–52), "Testimony of Edward Frazier," NA.
18. Tidwell, *April 65*, 202. Tidwell has reprinted the contents of the entire remnant on pages 198–204.
19. Ibid., 19–20.
20. Investigation and Trial Papers Relating to the Assassination of President Lincoln, Record Group M599, Reel 13, page 3815 (subpages 50–58), "Testimony of Edward Frazier," NA.
21. Letters Received by the Confederate Secretary of War, 1861–65, Record Group 109, Entry M437, Reel 124, Number 454-C-64, NA. The document is cosigned at the bottom by both Henry Dillingham and Thomas Clarke with a notation that reads, "We likewise vouch for their loyalty."
22. Letter from General Maury to the adjutant and inspector general discussing activities of Courtenay and McDaniel organizations, August 28, 1864, Letters Received by the Confederate Adjutant and Inspector General's Office, 1861–65, Record Group 109, Entry M474, Number 3972-B-64, NA.
23. Tidwell, *April 65*, 51. Tidwell cites the source of this document as "Company A, Secret Service, Compiled Service Records, RG 109." However, this file name is incorrect and the document's location at the National Archives is not known.

24. Hill, "Texan Gave World First Submarine Torpedo."
25. Thatcher, "The Courtenay Coal Torpedo," 5; Porter, "Torpedo Warfare," 225; Bradford, *History of Torpedo Warfare*, 62.
26. William Murphy states in testimony in February 1865 that Edward Frazier had planted a coal torpedo aboard a Federal supply vessel docked along the Mississippi River several months earlier. This information reinforces the theory that Frazier, Dillingham, and Clarke may have been issued such devices before leaving the Confederate capital. For additional information regarding issuing of coal torpedoes to Confederate political agents operating in Canada, see Mayers, "Spies across the Border," 61.
27. Investigation and Trial Papers Relating to the Assassination of President Lincoln, Record Group M599, Reel 13, page 3815 (subpage 53), "Testimony of Edward Frazier," NA.
28. Testimony of confessed Boat Burner William Murphy, February 22, 1865, "Baker Report on Boat Burners," Records of the Office of the Judge Advocate General (Army), Record Group 153, Miscellaneous Correspondence, Box 12, Entry 6, NA.
29. Ibid.
30. Ibid.
31. Ibid.
32. Ibid.
33. Ibid.
34. Porter, "Torpedo Warfare," 226. William Murphy, in testimony some months later, stated that Edward Frazier had been the saboteur responsible for planting the coal torpedo. This information only reinforces the theory that Frazier, Dillingham, and Clarke may well have been issued such devices before leaving the Confederate capital.
35. E. C. Singer letter, August 14, 1864, Manuscript Department, Brockenbrough Library, Museum of the Confederacy; photocopy filed in Hunley Archive, Warren Lasch Conservation Center.
36. *War of the Rebellion*, ser. 4, 3:802–4. For additional information regarding Leovy's activities, see Leovy Biography, Henry J. Leovy Papers, Williams Research Center; photocopy filed in Hunley Archive, Warren Lasch Conservation Center.
37. Leovy Biography, Henry J. Leovy Papers, Williams Research Center.
38. *War of the Rebellion*, ser. 4, 3:806.
39. Ibid., 802–4.
40. Letter from James McClintock to Matthew Fountain Maury, Matthew F. Maury Papers, vol. 46, item 9087–9094, Manuscript Division, Library of Congress.
41. *Scientific American* 2, no. 27 (March 27, 1847); 3, no. 52 (September 16, 1848); 4, no. 6 (October 28, 1848). Articles in *Scientific American* regarding electric motors increase in volume in the years just prior to the Civil War.
42. Letter from James McClintock to Matthew Fountain Maury, Matthew F. Maury Papers, vol. 46, item 9087–9094, Manuscript Division, Library of Congress.
43. Letters Received by the Confederate Secretary of War, 1861–65, Record Group 109, Entry M437, NA; photocopy filed in Hunley Archive, Warren Lasch Conservation Center.
44. Ibid.
45. Ibid.
46. Letter from Baxter Watson to General P. G. T. Beauregard, January 6, 1865, Hunley File, Mobile Historic Preservation Society; photocopy filed in Hunley Archive, Warren Lasch Conservation Center.

NOTES TO PAGES 144–150   217

47. Letters Sent, Department of Texas, New Mexico, and Arizona, September 1864–May 1865, Record Group 109, chap. 2, NA.
48. Entry dated November 11, 1864, regarding submitted report filed by E. C. Singer, Letters Received Engineer Bureau, Record Group 109, NA; photocopy of this handwritten entry filed in Hunley Archive, Warren Lasch Conservation Center.
49. Ibid.
50. Letters Received by the Confederate Adjutant and Inspector General's Office, 1861–65, Record Group 109, Entry M474, NA; photocopy filed in Hunley Archive, Warren Lasch Conservation Center.
51. Ibid.
52. Special Orders Issued by the Confederate Adjutant and Inspector General's Office, 1861–65, Record Group 109, NA.
53. *War of the Rebellion*, ser. 1, 42:1219–20.
54. Sleeman, *Torpedoes and Torpedo Warfare*, 1–4; Perry, *Infernal Machines*, 148.
55. *Official Records of the Union and Confederate Navies*, ser. 1, 11:160–62.
56. Perry, *Infernal Machines*, 151.
57. Sleeman, *Torpedoes and Torpedo Warfare*, 190.
58. Ibid.
59. Letter from General Walker dated December 2, 1864, Letters Sent, Department of Texas, New Mexico, and Arizona, September 1864–May 1865, Record Group 109, chap. 2, NA.
60. Special Order Number 101, December 20, 1864, Special Orders Department of Texas, New Mexico, and Arizona, September 1864–February 1865, Record Group 109, NA.
61. Summary of Robert Dunn letter and response, December 12, 1864, Endorsements, Department of Texas, New Mexico, and Arizona, Record Group 109, NA.
62. Testimonies of the two Confederate deserters Henry Jones and C. M. Hassenger: "The enemy are building a torpedo boat at Goose Creek, One hundred and Forty feet long to be clad in rail road iron." There is other testimony from this debriefing on March 27, 1865: "They [the deserters being questioned] know nothing definite about the torpedo boats, but have heard that such boats were being built on the San Jacinto River, at Lynchburg. One of the men saw what was shown to him as a torpedo boat lying in the main channel at Galveston. It was shaped like a box, with square corners, and was quite low in the water. He could not tell whether she was plated or not." It was not uncommon for the Confederates to use railroad iron to sheathe ironclads (the ironclad CSS *Missouri* constructed on the Red River is a good example). Navy Area File, Record Group 45, Entry 625, NA; photocopies of these handwritten documents filed in Hunley Archive, Warren Lasch Conservation Center.
63. Special Order Number 19, January 19, 1865, Special Orders, Department of Texas, New Mexico, and Arizona, September 1864–February 1865, Record Group 109, NA. The order reads, "Private J. A. Hemphill Co. 'G' Tannon's Regiment, being a carpenter first class, is hereby detailed for sixty (60) days to work in the Shipyard at Lubbock's Mill. He will report for duty without delay to Mr. J. D. Braman, Superintendent of Construction for Torpedo boat." This document is extremely important to the history of the Singer group, for it proves that Braman had been placed in charge of overseeing the construction of the second torpedo boat by early January 1865.
64. Relevant text reads, "Captain Thomas Chubb, Superintendent Steam boat repairs, Sir: You are instructed as follows: You will take charge of the dredge boat and at once proceed to dig out the mouth of Tyber Creek and canal to a depth sufficient to float vessels drawing five feet nine inches water. If this can be done before the machinery

is put into the torpedo boat, the latter will not be removed: If this cannot be done in time, you will remove the torpedo boat to the ship yard at Lynchburg, and hasten forward the work as rapidly as possible. The object of removing the torpedo boat to Lynchburg being to save it from being shut up in Tyber Creek, where it might be lost in case the enemy should get into the bay." Letter sent to Captain Chubb, February 21, 1865, Letters Sent, Department of Texas, New Mexico, and Arizona, September 1864–May 1865, Record Group 109, chap. 2, NA.

65. Ibid.
66. Letter from General Walker dated December 2, 1864, in ibid.
67. Report submitted by M. P. Hunnicutt, March 13, 1865, Letters Received, Office of the Chief Signal Officer, District of the Gulf and Mississippi, Box 4, Number 1407, NA.
68. Cotham, *Battle on the Bay*, 131.
69. Testimony of Confederate deserters Henry Jones and C. M. Hassenger, January 2, 1865, Navy Area File, Record Group 45, Entry M625, NA.
70. Ibid.
71. Hutchinson, *Life on the Texas Blockade*, 34–35.
72. Hunley File, Mobile Historic Preservation Society; photocopy filed in Hunley Archive, Warren Lasch Conservation Center.
73. Stern, *The Confederate Navy*, 238.
74. Flato, *The Civil War*, 195.
75. Report by M. P. Hunnicutt, March 13, 1865, Letters Received, Office of the Chief Signal Officer, District of the Gulf and Mississippi, Box 4, Number 1407, NA.
76. Ibid.
77. Letter from Henry Leovy to secretary of war, January 27, 1865, Letters Received by the Confederate Secretary of War, 1861–65, Record Group 109, Entry M437, Number 59-L-65, NA.
78. Ibid.
79. Letters Received by the Confederate Secretary of War, 1861–65, Record Group 109, Entry M437, Number 44-L-65, NA.
80. Letters Sent, Department of Texas, New Mexico, and Arizona, September 1864–May 1865, Record Group 109, chap. 2, NA.
81. Summary of Robert Dunn letter and response dated January 24, 1864, Endorsements, Department of Texas, New Mexico, and Arizona, Record Group 109, NA; photocopy filed in Hunley Archive, Warren Lasch Conservation Center.
82. *War of the Rebellion*, ser. 1, 47:1315.
83. Ibid., 1314.
84. Ibid., 1315.
85. Ibid.
86. Crumbley, *General Sherman's March*, 1.
87. Ibid., 2.
88. "Torpedo Warfare," by General Rains, Museum of the Confederacy, 6.
89. *War of the Rebellion*, ser. 1, vol. 48, pt. 1, 1197.
90. Report by M. P. Hunnicutt, March 13, 1865, Letters Received, Office of the Chief Signal Officer, District of the Gulf and Mississippi, Box 4, Number 1407, NA.
91. Ibid.
92. *Official Records of the Union and Confederate Navies*, ser. 1, 22:103–10.

93. *War of the Rebellion*, ser. 1, vol. 48, pt. 1, 1197.
94. *Official Records of the Union and Confederate Navies*, ser. 1, 26:184.
95. Ibid., 22:103–5.
96. Special Orders, Department of Texas, New Mexico, and Arizona, Record Group 109, NA.
97. Navy Area File, Record Group 45, Entry 625, NA.
98. The document proves that Edgar Singer had returned to Texas during the closing months of the war, and his official parole was executed at Port Lavaca in early July 1865. At the top of the document is written, "The undersigned Prisoner of War belongs to the Army of the Trans-Mississippi having been surrendered by General E. Kirby Smith, C.S.A." Unfiled Papers and Slips Belonging in Confederate Compiled Service Records, Record Group 109, Entry M347, NA; photocopy of this document and copies of both Dunn and Braman's parole papers filed in Hunley Archive, Warren Lasch Conservation Center.
99. A February 28, 1865, inventory of torpedoes then on hand in Mobile states that some eighty Singer mines were then available for deployment. Engineer Bureau Miscellaneous Records, Record Group 109, Box 1, NA; Perry, *Infernal Machines*, 185.
100. Perry, *Infernal Machines*, 182–85.
101. *Official Records of the Union and Confederate Navies*, ser. 1, 22:71.
102. Bradford, *History of Torpedo Warfare*, 46.
103. "Engineer Office, Mobile, August 5, 1864. Mr. McClintock, in charge of torpedoes, will proceed tonight with 30 torpedoes to or near the mouth of the Dog River.... By order of Col. Scheliha." Letters Sent by Engineering Department, District of the Gulf, Record Group 109, NA; photocopy filed in Hunley Archive, Warren Lasch Conservation Center.
104. Scharf, *History of the Confederate States Navy*, 594.

Chapter 6

1. Flato, *The Civil War*, 195. My great-great-grandfather's brother, Corp. Nathanial Pearce of the 24th North Carolina Regiment, was captured during the attack on Fort Stedman and taken to Point Lookout, where he remained a prisoner until the end of July. My great-great-grandmother's nephew, Lt. Alexander J. Bumpass (nicknamed "Tip" according to family records) of the same regiment, was killed in the assault. He had served in the 24th North Carolina since May 1861 and had seen action at the Seven Days, Antietam, Fredericksburg, and Petersburg.
2. *War of the Rebellion*, ser. 1, vol. 46, pt. 3, 1378.
3. Evans, *Judah P. Benjamin*, 294.
4. Hanna, *Flight into Oblivion*, 5.
5. Ibid.
6. Evans, *Judah P. Benjamin*, 295.
7. Perry, *Infernal Machines*, 138; Evans, *Judah P. Benjamin*, 294.
8. Evans, *Judah P. Benjamin*, 296.
9. Hanna, *Flight into Oblivion*, 11–12.
10. Information regarding George Trenholm's interest in purchasing the *Hunley* from the Singer group is mentioned in an August 29, 1863, letter written by Harriet Middleton (Trenholm at the time had not as yet been appointed to the post of Confederate States treasurer): "It seems a pity that the fish-boat should have been turned over to the Government, we might have had a better chance at the 'Ironsides' if she had

been bought by Trenholm and taken out by Jefferson Bennett, as was first proposed." Middleton Papers, South Carolina Historical Society.
11. Evans, *Judah P. Benjamin*, 296.
12. Hanna, *Flight into Oblivion*, 13.
13. Ibid., 15.
14. *War of the Rebellion*, ser. 1, vol. 46, pt. 3, 1383.
15. Special Orders, Department of Texas, New Mexico, and Arizona, Record Group 109, NA.
16. Letter Book of J. H. Carter, entry dated April 8, 1865, Commanding Naval Defenses in Western Louisiana, Record Group 109, NA; copy of this handwritten document filed in Hunley Archive, Warren Lasch Conservation Center.
17. Special Orders, Trans-Mississippi Department, 1864–65, chap. 2, vol. 79, NA.
18. Hanna, *Flight into Oblivion*, 76.
19. Mallory, "Last Days of the Confederate Government," 107.
20. Hanna, *Flight into Oblivion*, 6.
21. Evans, *Judah P. Benjamin*, 301.
22. Dowd, *Life of Zebulon B. Vance*, 485–86.
23. Ibid.
24. Hanna, *Flight into Oblivion*, 78.
25. *War of the Rebellion*, ser. 1, vol. 48, pt. 2, 1288–91.
26. Hanna, *Flight into Oblivion*, 46.
27. Tidwell, *April 65*, 31–35, 128.
28. Evans, *Judah P. Benjamin*, 303.
29. Ibid., 312; Hanna, *Flight into Oblivion*, 45.
30. It is known that Colonel Leovy must have traveled southward with this detachment because he is mentioned in an April 29, 1865, Abbeville dispatch addressed to President Davis. *War of the Rebellion*, ser. 1, vol. 49, pt. 2, p. 1269.
31. Hanna, *Flight into Oblivion*, 53.
32. Ibid., 58.
33. *War of the Rebellion*, ser. 1, vol. 49, pt. 2, 1269.
34. Evans, *Judah P. Benjamin*, 309–10.
35. Ibid.
36. *War of the Rebellion*, ser. 1, vol. 48, pt. 2, 1284.
37. Ibid.
38. Ibid.
39. Hanna, *Flight into Oblivion*, 81; *War of the Rebellion*, ser. 1, vol. 48, pt. 1, 298.
40. Special Orders, Department of Texas, New Mexico, and Arizona, March–May 1865, Record Group 109, NA.
41. Letters Sent, Department of Texas, New Mexico, and Arizona, September 1864–May 1865, Record Group 109, NA.
42. Letters Received, Department of Texas, New Mexico, and Arizona, Record Group 109, NA; copy of this handwritten document filed in Hunley Archive, Warren Lasch Conservation Center.
43. Ibid.
44. Telegrams Sent, "B District" Headquarters, Houston, July 28, 1864–May 3, 1865, Record Group 109, NA. Maj. Edward B. Hunt conducted self-propelled torpedo ex-

periments in early 1863 at the Brooklyn Navy Yard. Such devices, to be successful, had to displace their own weight in water. If the rocket-powered torpedo was too light, it would immediately veer toward the surface upon discharge; if too heavy, it would veer toward the bottom. To counteract these problems, Major Hunt cast his elongated, hollow torpedoes much like artillery shells. The end product, fired through a tube several feet beneath the water's surface, was about eight inches in diameter and three feet long and weighed about fifty pounds (the weight required to displace its volume in water). Captain Singer's boats were armed with this type of weaponry, so we can assume that the torpedoes were to be cast from molten copper. We know nothing more about how Captain Singer intended to utilize his self-propelled torpedoes.

45. Special Orders, Department of Texas, New Mexico, and Arizona, March–May 1865, Record Group 109, NA.
46. Hanna, *Flight into Oblivion*, 63.
47. "At Abbeville it was decided to reduce the train, and Colonel Wood, Lubbock and Johnston, and myself opened and destroyed many unimportant papers; the bulk of the remaining papers and stationery were repacked and left by order of Colonel Johnston with Mrs. Col. Henry J. Leovy at Abbeville." Letter from H. Clark to B. Harrison, February 20, 1866, Harrison Family Papers, Library of Congress; copy of this handwritten document filed in Hunley Archive, Warren Lasch Conservation Center.
48. Evans, *Judah P. Benjamin*, 312.
49. *War of the Rebellion*, ser. 1, vol. 49, pt. 2, 599.
50. Hanna, *Flight into Oblivion*, 195.
51. Evans, *Judah P. Benjamin*, 312.
52. Ibid., 309.
53. Ibid., 314.
54. "Henry J. Leovy, Speaker of the Day," *New Orleans Times-Democrat*.
55. Hanna, *Flight into Oblivion*, 108.
56. *War of the Rebellion*, ser. 1, vol. 49, pt. 1, 536–37.
57. Evans, *Judah P. Benjamin*, 314.
58. Ibid., 306.
59. Ibid., 314.
60. Ibid.
61. Ibid., 315.
62. Unfiled Papers and Slips Belonging in Confederate Compiled Service Records, Record Group 109, Entry M347, NA. Since Colonel Henry Leovy was not assigned to any specific unit, his May 21, 1865, parole document was placed into this file. Several documents signed by members of Captain Singer's unit are in this file because archivists in the last century had no information about to which organizations these men were attached. Singer, Dunn, and Braman's parole papers are also located in this file.
63. Letter from H. Clark to B. Harrison, February 20, 1866, Harrison Family Papers, Library of Congress.
64. We can only assume that Henry Leovy and his wife eventually turned the crates over to the Federal government, for Clark himself stated the documents were extremely important and likely to be of great value in recording the life of the Confederacy. Some weeks after parole Captain Clark returned to Abbeville and attempted to secure these archives. He wrote a letter concerning the reaction of Leovy's wife: "When I reached Abbeville I found Mrs. Leovy about starting for Louisiana. I stated my purpose, but she declined delivering me the papers, stating that she would deliver them

to Colonel Johnston only, or to his order, or the order of the president [Jefferson Davis].... She stated that she would carry the papers with her and deliver them to me whenever I produced a written order for them from Col. Johnston." Ibid.; copy of this handwritten document filed in Hunley Archive, Warren Lasch Conservation Center.

65. Special Orders, Department of Texas, New Mexico, and Arizona, March–May 1865, Record Group 109, NA.
66. "Your propositions for the surrender of the troops under my command are not such that my sense of duty and honor will permit me to accept." *War of the Rebellion*, ser. 1, vol. 48, pt. 1, 189.
67. Ibid., 193.
68. Ibid., 189–90.
69. Ibid.
70. Ibid.
71. Hunt, "The Battle of Palmito Ranch," 3.
72. *War of the Rebellion*, ser. 1, vol. 48, pt. 1, 267.
73. Ibid., 190–91.
74. Ibid., 192–94.
75. Ibid., 297–98.
76. Ibid., 298.
77. Hill, "Texan Gave World First Submarine Torpedo."
78. *War of the Rebellion*, ser. 1, vol. 48, pt. 2, page 1121.
79. Stuart, "First Submarine Torpedo Vessel."
80. A late-war Confederate document in the National Archives relates to the Houston-based submarine: "Headquarters Defenses Galveston, May 9, 1865. Order Number 125. There are torpedoes and fixtures belonging to submarine battery. The latter are detained under previous orders." Following this order appears a brief statement: "Shall ammunition be forwarded? Shall torpedoes and fixtures be forwarded? Very Respectfully, A. Smith, Colonel Commanding." *Webster's* defines fixture as "something firmly attached as a permanent part of some other thing." Thus, "fixtures belong to submarine battery" may have referred to machinery for the Houston-based submarine boat M. P. Hunnicutt mentioned in his report. Letters Received, Department of Texas, New Mexico, and Arizona, September 5, 1864–May 25, 1865, Record Group 109, Box 5, NA; copy of this handwritten document filed in Hunley Archive, Warren Lasch Conservation Center.
81. *War of the Rebellion*, ser. 1, vol. 48, pt. 1, 297–98. Just after World War I, my grandfather Robert E. Ragan, as a boy with two friends, found several crates of muskets buried under the floor of an old shed near where they lived in Hunt County, Texas. Local newspapers at the time reported that they were sure that the muskets had been secretly hidden there by disbanding Confederates at the end of the war so they could fight another day.
82. Ibid.
83. Parole papers of "E. C. Singer, John D. Braman, Robert W. Dunn," Unfiled Papers and Slips Belonging in Confederate Compiled Service Records, Record Group 109, Entry M347, NA; copies filed in Hunley Archive, Warren Lasch Conservation Center.

Conclusion

1. Maury, "First Marine Torpedoes Were Made in Richmond."
2. Ibid.

3. A scrapbook containing diagrams of all Confederate torpedoes recovered during the war years is filed in folder 43 of the Navy Subject File, Record Group 74, NA; several photocopies and attached documentation of various mines in this book filed in Hunley Archive, Warren Lasch Conservation Center.
4. Shela Fretwell has privately published a book on John R. Fretwell regarding his war service and postwar activities: *The Fretwell Family*, 16; copy of manuscript filed in Hunley Archive, Warren Lasch Conservation Center; "Fretwell Obituary," *Mobile Register*.
5. Fretwell, *The Fretwell Family*, 16; copy of manuscript filed in Hunley Archive, Warren Lasch Conservation Center; "Fretwell Obituary," *Mobile Register*.
6. Stuart, "First Submarine Torpedo Vessel."
7. Hill, "Texan Gave World First Successful Submarine Torpedo."
8. Census of 1860 and 1870, Calhoun County, Texas.
9. Ibid. Information regarding Frary's relationship to Edgar Singer is from a document in the author's collection (photocopy filed in Hunley Archive, Warren Lasch Conservation Center).
10. Information regarding the death of Baxter Watson is found in a document filed in the Louis Genella Collection, Tulane University; photocopies of all documents in this collection filed in Hunley Archive, Warren Lasch Conservation Center.
11. Henry J. Leovy Papers, Williams Research Center.
12. Ibid; "Henry J. Leovy, Obituary," *New Orleans Picayune*.
13. Tidwell, *April 65*, 106.
14. "Attempt to Fasten the Assassination of President Lincoln on President Davis"; Investigation and Trial Papers Relating to the Assassination of President Lincoln, Record Group M599, Reel 13, page 3815 (subpages 49–50), "Testimony of Edward Frazier," NA. Relevant testimony reads, "Question: Do you know anything of the burning of a hospital of the United States at Louisville? Answer: No sir, I do not. All I know about that is that there was a man put in jail, but I do not think he did it. Question: Do you know the name of the man who claimed compensation from the Confederate government for that service? What was the man's name? Answer: Dillingham. Question: What amount did he claim for burning that hospital? Answer: He did not claim any particular amount at Richmond. Question: When was the hospital burned? Answer: It must have been burned along June or July 1864. Question: Do you remember how the fire occurred? Was it at night and was it attended with any loss of life? Answer: It was at night. Question: You do not know whether there were any patients burned up or not? Answer: I believe there was nobody burned up; at least I never heard of anybody being burned.
15. Proof that Henry Dillingham was a member of the second *Hunley* crew is found in this order: "October 16th, 1863. Major: Give transportation to Lieutenant Dixon 21st Regiment Alabama Volunteers, and Henry Dillingham to Mobile and return, on business connected with the submarine torpedo boat. "Orders and Circulars, Department of South Carolina, Georgia, and Florida, Record Group 109, NA.
16. McClintock letter dated January 10, 1872, in "Submarine warfare, ADM 1/6236/99455," British Admiralty Archives; photocopy filed in Hunley Archive, Warren Lasch Conservation Center.
17. Matthew F. Maury Papers, vol. 46, item 9087–9094, Manuscript Division, Library of Congress.
18. "Submarine warfare, ADM 1/6236/99455," British Admiralty Archives; photocopies of all documents in this collection filed in Hunley Archive, Warren Lasch Conservation Center.

19. Description of McClintock, Eustace Williams Collection, Mobile Public Library.
20. William Alexander's 1903 speech before the Iberville Historical Society, in collection of the Mobile Public Museum; photocopy filed in Hunley Archive, Warren Lasch Conservation Center.
21. Mobile City Directory for 1872 lists William Alexander as being in partnership with Park and Lyons; photocopy filed in Hunley Archive, Warren Lasch Conservation Center.
22. "Builder of First Submarine Dead," *Mobile Register*.
23. George E. Dixon Estate File, Probate Court, Mobile County, Alabama; photocopies of all relevant documents from this collection filed in Hunley Archive, Warren Lasch Conservation Center. The list of articles in Dixon's leather trunk that were to be auctioned includes "1 Lieutenant's Confederate Uniform, 7 yards gray cashmere [probably to make additional uniforms], 1 Fine Black cloth coat and pants, 3 pair cashmere pants, 1 Flannel coat, 7 vests, 1 White linen suit—coat, pants and vest, 2 Striped linen coats, 2 pair gray flannel drawers, 3 linen under shirts, 6 cloth shirts, 1 box Masonic books, 1 leghorn hat, 1 diamond pin."
24. Parole papers of Robert W. Dunn, Unfiled Papers and Slips Belonging in Confederate Compiled Service Records, Record Group 109, Entry M347, NA; Texas State Census of 1860, 1870, and 1880.
25. Hill, "Texan Gave World First Successful Submarine Torpedo."
26. King, *Torpedoes*, 12.
27. "Torpedo Inventor Dies: Death of Captain Edgar Collins Singer Occurs at Home in Marlin, Texas." A photocopy of this obituary filed in Hunley Archive, Warren Lasch Conservation Center.
28. Tidwell, *April 65*, 78.
29. Ibid.
30. Much has been written regarding Booth's assumed relationship with the Confederate Secret Service. See ibid.
31. In early April 1865, Secretary of State Judah P. Benjamin ordered the destruction of all Secret Service files in Richmond. With this act, most of the history regarding that unique organization was lost forever.
32. A thorough search of the index to "Miscellaneous Letters Received by the Secretary of the Navy" (all nineteenth-century letters sent by civilians to the Navy Department were lumped into this category) revealed that none of the Singer group contacted the navy to share torpedo-related information between 1865 and 1870, so they most likely never did.
33. "Since we have been on this side of the river we have gotten up a great many projects and have been interested in many new schemes, the particulars of which are too lengthy for an ordinary letter. Among the number, however, was a submarine boat, built at this place, of which Whitney and myself bought one-fifth for $3,000." If John Braman had written just a few lines regarding these "great many projects," we would today have a much better understanding of the Singer group's covert activities. Letter from John D. Braman to his wife, March 3, 1864, Confederate Navy Subject File, Reel 11, Record Group 109, NA.
34. King, *Torpedoes*, 12.
35. Hill, "Texan Gave World First Successful Submarine Torpedo."
36. Housatonic "Z" File, Naval Archives, Washington Navy Yard.
37. After years of studying various Confederate records, I have yet to find any mention of a bounty actually being paid by the Confederate government for the destruction

of a Federal vessel. If the Singer group were in fact not being paid, it seems strange that they would have continued in such a dangerous occupation, for it was continually rumored that Confederate torpedo operators, wherever captured by the Federals, were to be immediately executed.

38. The Masonic Creed reads as follows: "A Mason must be a good man before he can enter the Fraternity. By becoming a Mason, he strives to become a better man. A Mason must be true to himself and more importantly, he must be true to his fellow man. A Mason must have a belief in a Supreme Being before he can become a Mason. No Atheist can ever be made a Mason. It is more blessed to give than to receive. Masons are involved in helping others. Masons act with honor in everything they do. Masonry insists on toleration, on the right of each person to think for himself in religion, social, and political matters."

39. Mobile historian and resident Charles Stalmach originally brought my attention to these documents.

40. William Alexander was buried with Masonic honors in Mobile's Magnolia Cemetery in 1914 (where John Fretwell was buried in 1885). Information regarding his Masonic history and involvement with the *Hunley* submarine was distributed to the mourners in a two-page document, a copy of which is now in the Hunley File, Mobile Historical Preservation Society; photocopy filed in Hunley Archive, Warren Lasch Conservation Center.

41. After three expeditions at a cost exceeding well over $130,000, Clive Cussler's National Underwater and Marine Agency (NUMA) dive team (Ralph Wilbanks, Wes Hall, and Harry Pecorelli) discovered the intact *Hunley* on May 3, 1995. Cussler is best known as an adventure novelist whose international book sales have topped over one hundred million. The Confederate submarine *H. L. Hunley* is only one of many historic vessels that NUMA has searched for and discovered.

Bibliography

MANUSCRIPT COLLECTIONS

Atlanta Federal Archives and Records Center, East Point, Georgia
Record Group 77, Miscellaneous Wrecks, 1871–88

Author's Collection
Letter written by Isaac Ball regarding Civil War activities of Reverend Franklin G. Smith (July 26, 1919)
"The Sinking of the CSS *Hunley*, James McClintock's Secret Journey to Halifax and His Influence on the Royal Navy's British Submarine Policy" (a paper written by Peter Hore, n.d.)

British Admiralty Archives
Submarine warfare, ADM 1/6236/99455

Collection of Dr. Charles Perry, Charleston, South Carolina
"Brother Charles Hazelwood Hasker and the Fish," H. Metzger
Letter from James Tomb, June 7, 1908

Georgia Historical Society, Savannah, Georgia
My Dear Friends on Morris Island, Collection #19

Hill Memorial Library, Louisiana State University, Baton Rouge
Marshall Furman Papers

Library of Congress, Washington, DC
Washington Irving Chambers Papers, "Confederate Torpedoes"
Confederate States Records, Register of Commissions
Furman, Green, and Chandler Papers
Harrison Family Papers
Matthew F. Maury Papers
E. Willis Scrapbook, "Torpedoes and Torpedo Boats"

Mobile City Museum
Hunley File

Mobile Historical Preservation Society, Mobile, Alabama
Hunley File

Mobile Public Library, Mobile, Alabama
Eustace Williams Collection

Museum of the Confederacy, Eleanor S. Brockenbrough Library, Richmond, Virginia
Alphabetical List of Patentees for the Year 1863
Annual Reports of the Confederate Commissioner of Patents, 1861–65

BIBLIOGRAPHY

Confederate Navy, "The First Submarine Boats"; "Sinking of the Housatonic by One"; "Sad Fate of Former Crew"; "Reminiscences of a Confederate Naval Officer"; filed in "North Carolina Leaf"
Letters Sent by the Confederate Commissioner of Patents 1863
"Torpedo Warfare" by General Rains

National Archives, Washington, DC
Account Book of the Auditor's Office, Trans-Mississippi Department
P. G. T. Beauregard Private Papers
Book of Endorsements, District of Texas, New Mexico, and Arizona, April–August 1864
Census of Calhoun County, Texas, 1860
Charleston Station Payrolls and Shipping Articles, Entry T-829
Compiled Service Records of Confederate Generals and Staff Officers, and Non-Regimental Enlisted Men
Compiled Service Records of Confederate Soldiers Who Served in Organizations from the State of Alabama
Compiled Service Records of Confederate Soldiers Who Served in Organizations from the State of South Carolina
Compiled Service Records of Soldiers Who Served in Organizations from the State of Texas
Confederate Navy Subject File
Confederate Papers Relating to Citizens or Business Firms
Confidential Letters and Telegrams Sent, Trans-Mississippi Department, 1865
Endorsements, Department of Texas, New Mexico, and Arizona
Endorsements, Telegrams Sent, Department of South Carolina, Georgia, and Florida
Endorsements on Letters Received, Department of Alabama, Mississippi, and Louisiana, November 1863–May 1865
Endorsements on Letters Received, Department of South Carolina, Georgia, and Florida, November 1862–February 1864
Endorsements on Letters Received, General W. H. C. Whiting's Command, Wilmington, North Carolina, 1863–64
General and Special Orders of Sub-commands, Department of South Carolina, Georgia, and Florida, 1863–64
General Orders, Department of South Carolina, Georgia, and Florida, July 1862–January 1864
General Orders, Department of Texas, New Mexico, and Arizona
Gillmore Papers (Department of the South)
Captain M. M. Gray's Letter Book
Index to the Letters Received by the Confederate Adjutant and Inspector General and by the Confederate Quartermaster General, 1861–65
Index to the Letters Received by the Confederate Engineering Department
Index to the Letters Received by the Confederate Secretary of War, 1861–65
Index to Letters Received, Department of South Carolina, Georgia, and Florida
Inspection Reports, Department of South Carolina, Georgia, and Florida, 1863–64
Investigation and Trial Papers Relating to the Assassination of President Lincoln
Journal of Operation in Charleston Harbor
Letter Book of J. H. Carter
Letters and Telegrams Received, Department of Alabama, Mississippi, and East Louisiana, 1862–65

Letters and Telegrams Received, Department of the West
Letters and Telegrams Received, Trans-Mississippi Department
Letters and Telegrams Sent, Engineer Office at Charleston
Letters and Telegrams Sent, Engineer Office at Mobile, June–July 1863
Letters and Telegrams Sent, General P. G. T. Beauregard's Command
Letters and Telegrams Sent, Signal Corps, Department of South Carolina, Georgia, and Florida
Letters and Telegrams Sent and Endorsements on Letters Received by the Engineer Office, Department of South Carolina, Georgia, and Florida, 1863–64
Letters Received, Department of South Carolina, Georgia, and Florida, 1862–64
Letters Received, Department of Texas, New Mexico, and Arizona, December 31, 1862–April 30, 1863
Letters Received by the Confederate Adjutant and Inspector General, 1861–65
Letters Received by the Confederate Secretary of War, 1861–65
Letters Received by the Navy Department
Letters Received by Office of the Chief Signal Officer, District of the Gulf and Mississippi
Letters Received by the Secretary of the Navy from Commanding Officers of Squadrons, 1841–86
Letters Received by the Secretary of the Navy from Officers below the Rank of Commander, 1802–84
Letters Received by the Secretary of the Navy: Miscellaneous Letters, 1801–84
Letters Referred to the Permanent Commission, Record Group 45, Entry 363
Letters Sent, Department of Alabama, Mississippi, and Louisiana, 1864–65
Letters Sent, Department of South Carolina, Georgia, and Florida, July 1862–April 1864
Letters Sent, Department of South Carolina, Georgia, and Florida, 1863–64
Letters Sent, Department of Texas, New Mexico, and Arizona
Letters Sent, Department of the Trans-Mississippi
Letters Sent, District of West Louisiana
Letters Sent, Engineer Office, District of the Gulf, 1863–65
Letters Sent, Engineer Office at Mobile
Letters Sent, General W. H. C. Whiting's Command, Wilmington, North Carolina
Letters Sent by the Chief of Artillery, Department of Alabama, Mississippi, and Louisiana, 1864–65
Letters Sent by Confederate Engineering Department
Letters Sent by the Engineer Office, Department of Alabama, Mississippi, and East Louisiana
Letters Sent by the Secretary of the Navy to Officers, 1798–1868
Letters Sent by the Signal Office, Department of South Carolina, Georgia, and Florida
Letters, Telegrams, and Orders, Department of South Carolina, Georgia, and Florida, March 1864
Miscellaneous Letters Sent by the Secretary of the Navy, 1798–1886
Miscellaneous Papers, Department of South Carolina, Georgia, and Florida, 1861–62, 1864
Miscellaneous Papers of the Confederate Engineering Bureau
Navy Area File
Navy Department Bureau Letters, September–December 1861
Navy Subject File, Record Group 45, Entry "AV"—Alligator File, Box 119
Office of the Chief Signal Officer, District of the Gulf and Western Mississippi

Order Book, Department of Texas, New Mexico, and Arizona
Orders, District of the Gulf, 1862–65
Orders, Eastern Sub-district of Texas and the Post of Austin, Texas, 1863–64
Orders and Circulars, Department of Alabama, Mississippi, and Louisiana, 1862–65
Orders and Circulars, Department of South Carolina, Georgia, and Florida, September 1863–March 1864
Orders and Circulars, Department of Texas, New Mexico, and Arizona
Orders and Circulars, Department of the West
Orders and Circulars, 1st Military District, Department of South Carolina, Georgia, and Florida, 1862–63
Orders and Circulars, Trans-Mississippi Department
Papers Pertaining to Vessels of or Involved with the Confederate States of America, "Vessel Papers"
Papers Relating to Confederate War Department Accounts
Proceedings of the Naval Court of Inquiry, Case Number 4345, February 26, 1864
Receipts for General Orders, Department of South Carolina, Georgia, and Florida
Record of Details, Department of Texas, New Mexico, and Arizona
Record of Persons and Articles Hired by Major E. G. Mohler, Quartermaster at Mobile
Records of Civilian Employment, Department of South Carolina, Georgia, and Florida, 1863
Records of the Office of the Judge Advocate General (Army), Correspondence, Letters Received
Records of the Passport Office at Richmond, 1861–65
Records Relating to Confederate Naval and Marine Personnel
Register of Letters and Telegrams Received, Department of Alabama, Mississippi, and Louisiana, 1862–65
Registers of Letters Received, Department of South Carolina, Georgia, and Florida, October 1862–November 1864
Reports, Department of South Carolina, Georgia, and Florida, 1863–64
Requests for Funds, Confederate War Department
Rolls of Civilians, Negroes, and Detailed Men Employed by Major Klumph, Quartermaster at Mobile
Roster of Commissioned Officers in the Eastern Sub-district of Texas
Special Orders, Department of Texas, New Mexico, and Arizona
Special Orders, Department of the Trans-Mississippi
Special Orders, Department of the Trans-Mississippi, 1864, Received by the District of Texas
Special Orders, Eastern Sub-district of Texas, New Mexico, and Arizona
Special Orders and Circulars, Department of South Carolina, Georgia, and Florida, September 1862–December 1863
Special Orders Issued by the Confederate Adjutant and Inspector General's Office, 1861–65
Telegrams Received, Department of Texas, New Mexico, and Arizona
Telegrams Sent, "B District" Headquarters, Houston, Texas
Telegrams Sent, Department of Alabama, Mississippi, and Louisiana, January 1864–April 1865
Telegrams Sent and Orders, Department of South Carolina, Georgia, and Florida, 1863–64
Telegrams Sent by General W. B. Taliaferro
Telegraphic Dispatches, Charleston, 1863
Unfiled Papers and Slips Belonging in Confederate Compiled Service Records

Union Provost Marshal's File, St. Louis, Missouri

Naval Archives, Washington Navy Yard, Washington, DC
Letters Received (Hunley File)

Probate Court, Mobile County, Alabama
George E. Dixon Estate File

South Carolina Historical Society, Charleston, South Carolina
Middleton Papers
Augustine Smyth, Communication from 1890s
Smyth Papers

South Carolinian Library, Columbia, South Carolina
Papers of Theodore A. Honour

Southern Historical Society Collection, University of North Carolina, Chapel Hill
Franklin Buchanan Letter Books
Ellen Shackelford Gift Papers
James Tomb Papers

St. Tammany Parish Court House Records, St. Tammany Parish, Louisiana
Will of H. L. Hunley

Tulane University, Manuscript Division, New Orleans, Louisiana
Louis Gennella Collection

Valentine Museum, Richmond, Virginia
Conrad Wise Chapman Collection

Virginia State Library, Richmond, Virginia
Tredegar Iron Works Record Books, October 1861

Warren Lasch Conservation Center, North Charleston, South Carolina
Hunley Archive

Williams Research Center, New Orleans, Louisiana
Henry J. Leovy Papers

PUBLICATIONS
Arnold, Thomas Jackson. *Early Life and Letters of Thomas J. Jackson*. New York: Fleming H. Revell, 1916.
Bakeless, John. *Spies of the Confederacy*. Philadelphia: J. B. Lippincott, 1970.
Bakeless, Katherine. *Confederate Spy Stories*. Philadelphia: J. B. Lippincott, 1973.
Barnes, John. *Submarine Warfare*. New York: Van Nostrand, 1869.
Bergeron, Arthur. *Confederate Mobile*. Jackson: University Press of Mississippi, 1991.
The Blockade. Alexandria, VA: Time-Life Books, 1983.
Bradford, R. B. *History of Torpedo Warfare*. Newport, 1882.
Brewer, Willis. *Alabama: Her History, Resources, War Record, and Public Men*. Montgomery, AL: Barrett & Brown, 1872.
Bucknill, John. *Submarine Mines and Torpedoes As Applied to Harbor Defense*. New York: Wiley, 1889.
Bulloch, James. *The Secret Service of the Confederate States in Europe*. New York: T. Yoseloff, 1959.
Burton, E. Milby. *The Siege of Charleston*. Columbia: University of South Carolina Press, 1970.

Busby, F. *Manned Submersibles.* Washington, DC: Office of the Oceanographer of the Navy, 1976.
Cable, George. *Strange True Stories of Louisiana.* St. Clair Shores, MI: Scholarly Press, 1889.
Cardozo, J. N. *Reminiscences of Charleston.* Charleston, SC: Joseph Walker, 1866.
Civil War Naval Chronology, 1861–1865. Washington, DC: Government Printing Office, 1971.
Coker, P. C. III. *Charleston's Maritime Heritage 1670–1865.* Charleston, SC: Coker-Craft Press, 1987.
Coski, J. *Capitol Navy: The Men, Ships and Operations of the James River Squadron.* Mason City, IA: Savas Publishing, 1996.
Cotham, Edward T. *Battle on the Bay: The Civil War Struggle for Galveston.* Austin: University of Texas Press, 1998.
Craighead, Edwin. "The *Hunley.*" In *Mobile's Past: Sketches of Memorable People and Events.* Mobile, AL: Powers Printing, 1925.
Crumbly, Tony. *General Sherman's March across North Carolina.* Fayetteville, NC: Carolina Coin & Stamp, 2000.
Current, Richard. *Encyclopedia of the Confederacy.* New York: Simon & Schuster, 1993.
Davis, B. *The Civil War: Strange and Fascinating Facts.* New York: Fairfax Press, 1982.
Davis, William. *An Honorable Defeat: The Last Days of the Confederate Government.* New York: Harcourt Publishing, 2001.
DeKay, J. *The Battle of Stonington: Torpedoes, Submarines, and Rockets in the War of 1812.* Annapolis, MD: Naval Institute Press, 1990.
Delaney, Caldwell. *Confederate Mobile: A Pictorial History.* Mobile, AL: Haunted Book Shop, 1971.
Dorset, Phyllis Flanders. "C.S.S. *Pioneer.*" In *Historic Ships Afloat.* New York: Macmillan, 1967.
Dowd, C. *The Life of Zebulon B. Vance.* Charlotte, NC: Observer Printing and Publishing House, 1897.
Duncan, Ruth Henley. *The Captain and Submarine, C.S.S. H. L. Hunley.* Memphis, TN: S. C. Toof, 1965.
Durkin, J. T. *Stephen R. Mallory: Confederate Navy Chief.* Chapel Hill: University of North Carolina Press, 1954.
Earle, A., and A. Giddings. *Exploring the Deep Frontier.* Washington, DC: National Geographic Society, 1980.
Ehrenkrook, F. *History of Submarine Mining and Torpedoes*: Berlin, 1878.
Evans, Eli. *Judah P. Benjamin, the Jewish Confederate*: New York: Free Press, 1988.
Field, Cyril. *The Story of the Submarine from the Earliest Ages to the Present Day.* London: Sampson Low, Martin, 1908.
Flato, Charles. *The Civil War.* New York: Golden Press, 1961.
Folmar, John Kent. *From That Terrible Field: Civil War Letters of James M. Williams, Twenty-First Alabama Infantry Volunteers.* Tuscaloosa: University of Alabama Press, 1981.
Fulton, Robert. *Torpedo War and Submarine Explosions.* New York: W. Eliot, 1810.
Fyfe, Herbert C. *Submarine Warfare: Past and Present.* New York: E. P. Dutton, 1907.
Gray, Edwin. *The Devil's Device: Robert Whitehead and the History of the Torpedo.* Annapolis, MD: Naval Institute Press, 1991.
Gunston, Bill. *Submarines.* Poole, UK: Blandford Press, 1976.
Hanna, A. J. *Flight into Oblivion.* Bloomington: Indiana University Press, 1938.
Harris, P. *Great Lakes First Submarine.* Michigan City: Michigan City Historical Society, 1982.

Henningsen, Charles. *Torpedoes: Memorial Respectfully Addressed to the Congress of the Confederate States.* Richmond, 1862.

Hoehling, A. A. *Damn the Torpedoes! Naval Incidents of the Civil War.* Winston-Salem, NC: John F. Blair Publisher, 1989.

Holmes, Emma. *The Diary of Miss Emma Holmes 1861–1866.* Baton Rouge: Louisiana State University Press, 1979.

Horton, E. *The Illustrated History of the Submarine.* London: Sidgwick & Jackson, 1974.

Hoyt, Edwin Palmer. *From the Turtle to the Nautilus: The Story of Submarines.* Boston: Little, Brown, 1963.

Hunt, Jeffrey William. "Battle of Palmito Ranch." In *Handbook of Texas Online.* Austin: Texas State Historical Association, 1997. Available at https://www.tshaonline.org/hand book/online/articles/qfp01.

Hutchinson, William. *Life on the Texas Blockade.* Providence, RI: The Society, 1883.

Johnson, John. *The Defense of Charleston Harbor.* Charleston, SC: Walker, Evans & Cogswell, 1890.

Keatts, C., and G. Farr. *U.S. Submarines.* Houston: Pisces Books, 1991.

Kerby, Robert. *Kirby Smith's Confederacy: The Trans-Mississippi South 1863–1865.* Tuscaloosa: University of Alabama Press, 1972.

Kinch, Oscar. *Confederate Operations in Canada and the North.* North Quincy, MA: Christopher Publishing House, 1970.

King, W. *Torpedoes: Their Invention and Use, from the First Application to the Art of War to the Present Time.* Washington, DC, 1866.

Kloeppel, James E. *Danger beneath the Waves: A History of the Confederate Submarine* H. L. Hunley. Orangeburg, SC: Sandlapper Publishing, 1987.

Lake, Simon. *The Submarine in War and Peace: Its Development and Its Possibilities.* Philadelphia: J. B. Lippincott, 1918.

Lawliss, C. *The Submarine Book.* New York: Thames & Hudson, 1991.

Lipscomb, Frank Woodgate. *Historic Submarines.* London: Hugh Evelyn, 1970.

Manigault, Edward. *Siege Train: The Journal of a Confederate Artilleryman in the Defense of Charleston.* Columbia: University of South Carolina Press, 1986.

Markle, Donald. *Spies and Spy Masters of the Civil War.* New York: Barnes & Noble Books, 1995

Martin, Fredrick. *Submarine Mining and Torpedoes.* Willetts Point, NY: Battalion Press, 1879.

Maury, R. *A Brief Sketch of the Work of Matthew Fontaine Maury during the War 1861–1865.* Richmond, VA: Whittet & Shepperson, 1915.

McCown, Leonard. *Calhoun County, Texas: Tenth Census of the United States, 1880.* Irving, TX: McCown, 1978.

Melton, Maurice. *The Confederate Ironclads.* South Brunswick, NJ: Thomas Yoseloff, 1968.

Middleton, D. *Submarine: The Ultimate Naval Weapon—Its Past, Present & Future.* Chicago: Playboy Press, 1976.

Miller, Francis Trevel. *The Photographic History of the Civil War.* New York: Thomas Yoseloff, 1957.

Moebs, Thomas Truxton. *Confederate States Navy Research Guide.* Williamsburg, VA: Moebs, 1991.

Nichols, James Lynn. *Confederate Engineers.* Tuscaloosa, AL: Confederate Publishing, 1957.

Norlin, F. E. *A Short History of Undersea Craft.* Newport, RI: Naval Torpedo Station, 1960.

Official Records of the Union and Confederate Navies in the War of the Rebellion. Washington, DC: Government Printing Office, 1901.

Ortzen, Len. *Stories of Famous Submarines*. London: Redwood Press, 1973.
Orvin, Maxwell Clayton. *In South Carolina Waters 1861–1865*. Charleston, SC: Southern Publishing, 1961.
Parker, William Hawar. *The Confederate States Navy*. Atlanta, 1899.
Perry, Milton F. *Infernal Machines: The Story of Confederate Submarine and Mine Warfare*. Baton Rouge: Louisiana State University Press, 1965.
Pinkerton, A. *The Spy of the Rebellion: A True History of the Spy System of the United States Army during the Late Rebellion*. Hartford, CT: M. A. Winter, 1883.
Porter, David D. "The Opening of the Lower Mississippi." In *Battles and Leaders of the Civil War: Grant-Lee Edition*. New York: Century, 1887.
Pratt, Fletcher. *The Civil War on Western Waters*. New York: Holt Publishing, 1956.
Robinson, William Morrison, Jr. *The Confederate Privateers*. New Haven, CT: Yale University Press, 1928.
Roland, A. *Underwater Warfare in the Age of the Sail*. Bloomington: Indiana University Press, 1978.
Roman, Alfred. *The Military Operations of General Beauregard in the War between the States, 1861–1865*. New York: Harper & Brothers, 1884.
Roscoe, Theodore. *Picture History of the U.S. Navy from Old Navy to New, 1776–1897*. New York: Charles Scribner's Sons, 1956.
Roscoe, Theodore, and Fred Freeman. *Picture History of the U.S. Navy*. London: Charles Scribner's Sons, 1956.
Rule, G. E. *Joseph W. Tucker and the Boat Burners*. St. Louis, MO: G. E. Rule, 2001.
Schafer, Louis S. *Confederate Underwater Warfare*. Jefferson, NC: McFarland, 1996.
Scharf, J. Thomas. *History of the Confederate States Navy*. Albany, NY: Joseph McDonough, 1894.
Selfridge, T. *What Finer Tradition: The Memoirs of Thomas O. Selfridge, Jr., Rear Admiral, U.S.N.* Columbia: University of South Carolina Press, 1987.
Sherman, Roger. *Joseph Henry's Contributions to the Electromagnetic and the Electric Motor*. Washington, DC: Smithsonian Institution Press, 1999.
Sleeman, C. W. *Torpedoes and Torpedo Warfare*. Portsmouth, 1880.
Solomon, Robert S. *The C.S.S. David: The Story of the First Successful Torpedo Boat*. Columbia, SC: R. L. Bryan, 1970.
Stern, Philip Van Doren. *The Confederate Navy: A Pictorial History*. Garden City, NY: Doubleday, 1962.
Still, William. *The Confederate Navy*. Annapolis, MD: Naval Institute Press, 1997.
———. *Confederate Shipbuilding*. Columbia: University of South Carolina Press, 1987.
Sueter, Murray. *The Evolution of the Submarine Boat, Mine and Torpedo, from the Sixteenth Century to the Present Time*. Portsmouth, UK: J. Griffin, 1907.
Tidwell, W. *April 65: Confederate Covert Action in the American Civil War*. Kent, OH: Kent State University Press, 1995.
Von Scheliha, V. *A Treatise on Coast-Defence*. London: E & F. N. Spon Publisher, 1868.
Wakelyn, Jon. *Biographical Directory of the Confederacy*. Westport, CT: Greenwood Press, 1977.
War of the Rebellion: A Compilation of the Official Records of the Union and Confederate Armies. Washington, DC: Government Printing Office, 1901.
Warner, Ezra J. *Generals in Gray*. Baton Rouge: Louisiana State University Press, 1959.
Watson, William J. *Adventures of a Blockade Runner, or Trade in Time of War*. New York, 1892.
Way, Frederick Jr., comp. *Way's Packet Directory 1848–1983*. Athens: Ohio University Press, 1983.

Wells, Thomas Henderson. *The Confederate Navy: A Study in Organization*. Tuscaloosa: University of Alabama Press, 1971.

Wideman, John. *The Sinking of the USS Cairo*. Jackson: University Press of Mississippi, 1993.

PERIODICALS

Alexander, William A. "The Confederate Submarine Torpedo Boat Hunley." *Gulf States Historical Magazine* (September 1902).

———. "The Heroes of the *Hunley*." *Munsey Magazine*, August 1903.

———. "Thrilling Chapter in the History of the Confederate States Navy." *Southern Historical Society Papers* (1902).

Arthur, S. C. "*Pioneer*: The First Submarine Boat Now on Exhibition in Jackson Square, New Orleans." *Alabama Historical Quarterly* (Fall 1947).

"The Attempt to Fasten the Assassination of President Lincoln on President Davis and Other Innocent Parties." *Southern Historical Society Papers* 9 (1881): 313.

"Bag for Lifting Wrecks." *Scientific American*, March 13, 1858.

Baird, G. W. "Submarine Torpedo Boats." *Journal of American Societies of Naval Engineers* 14, no. 2 (1902).

Balasses, George. "Dixie's Most Daring Sub." *Challenge Magazine*, August 1959.

Beard, W. E. "The Log of the C. S. Submarine." *U.S. Naval Institute Proceedings* (1916).

Beauregard, P. G. T. "Torpedo Service in the Harbor and Water Defense of Charleston." *Southern Historical Society Papers* (April 1878).

Blair, C. H. "Submarines of the Confederate Navy." *U.S. Naval Institute Proceedings* (October 1952).

"Boiler Explosion on the *Chenango*." *Scientific American*, April 30, 1864.

Bowman, B. "The *Hunley*: Ill-Fated Confederate Submarine." *Civil War History* (September 1959).

Brown, H. D. "The First Successful Torpedo and What It Did." *Confederate Veteran* 18 (1910): 169.

Brown, I. "Confederate Torpedoes in the Yazoo." In *Battles and Leaders of the Civil War* 3 (1888).

Crowley, A. "The Confederate Torpedo Service." *Century Magazine*, June 1898.

Davidson, H. "Davis and Davidson. A Chapter of War History concerning Torpedoes." *Southern Historical Society Papers* 24 (1896): 284–90.

———. "The Electrical Submarine Mine." *Confederate Veteran* (November 1908).

———. "Electric Torpedoes as a System of Defense." *Southern Historical Society Papers* 2 (1876).

"Demijohn Torpedoes." *Scientific American*, January 10, 1863.

"Destruction of a United States Steamer by a Torpedo." *Scientific American*, March 12, 1864.

Doran, C. "First Submarine in Actual Warfare." *Confederate Veteran* (1908).

"Explosion of a Submarine Torpedo." *Scientific American*, January 3, 1863.

"Fighting *Hunley*. Made in Mobile." *Port of Mobile*, December 1973.

Fitzhugh, Lester N. "Saluria, Fort Esperanza, and Military Operations on the Texas Coast, 1861–1864." *Southwestern Historical Quarterly* 61, no. 1 (1957): 66–10.

Ford, A. P. "The First Submarine Boat." *Confederate Veteran* (November 1908).

Fort, W. B. "The First Submarine in the Confederate Navy." *Confederate Veteran* 26, no. 10 (October 1918).

Franklin, Robert M. "Battle of Galveston." *U.C.V. Magazine*, April 2, 1911.

Glassel, W. T. "Reminiscences of Torpedo Service in Charleston Harbor." *Southern Historical Society Papers* 4 (1887): 225.
Hagerman, G. "Confederate Submarines." *U.S. Naval Institute Proceedings* (September 1977).
Hanks, C. C. "They Called Her a Coffin." *Our Navy* (March 1944).
Harrison, Burton. "The Capture of Jefferson Davis." *Century Magazine*, November 1883.
Kelln, A. L. "Confederate Submarines." *Virginia Magazine of History and Biography*, July 1953.
———. "Confederate Submarines and PT Boats." *All Hands* (April 1956).
Levy, Gordon S. "Torpedo Boat at Louisiana Soldiers' Home." *Confederate Veteran* 17, no. 9 (September 1909): 459.
"Loss of the *Housatonic*." Correspondence Section, *Army and Navy Journal* (March 5, 1864).
Mallory, S. "Last Days of the Confederate Government." *Southern Historical Society Papers* 15 (1887).
Maury, Dabney. "How the Confederacy Changed Naval Warfare: Ironclads and Torpedoes." *Southern Historical Society Papers* 22 (1894): 75.
———. "Remarkable History of a Torpedo Boat." *Army and Navy Journal* (August 25, 1866).
Mayers, Adam. "Spies across the Border." *Civil War Times* (June 2001).
Mazet, Horace S. "Tragedy and the Confederate Submarines." *U.S. Naval Institute Proceedings* (May 1942).
Morris, D. R. "The Rebels and the Pig Boat." *Argosy*, October 1954.
"A New Rebel Torpedo." *Scientific American*, December 19, 1863.
"New Rebel Torpedo." *Scientific American*, October 29, 1864.
"The New Torpedo in Charleston." *Scientific American*, September 12, 1863.
Olmstead, Charles H. "Reminiscences of Service in Charleston Harbor in 1863." *Southern Historical Society Papers* (1883).
"Organization of a Torpedo Corps." *Army and Navy Journal* (July 31, 1869).
Porter, D. D. "Torpedo Warfare." *North American Review* 27, no. 264 (September–October 1878).
Powles, J. M. "*Hunley* Sinks the *Housatonic*!!" *Navy Magazine*, January 1965.
Rains, G. "Torpedoes." *Southern Historical Society Papers* 3 (1877): 255.
"A Rebel Infernal Machine." *Scientific American*, January 10, 1863.
"The Rebel Torpedoes." *Scientific American*, June 18, 1864.
Schell, S. H. "Submarine Weapons Tested at Mobile during the Civil War." *Alabama Review* (July 1992).
Selfridge, Thomas O. "Retreat with Honor." *Battles and Leaders of the Civil War* 4 (1884): 362.
Shugg, W. "Profit of the Deep: The *H. L. Hunley*." *Civil War Times Illustrated* (1973).
Sims, L. "The Submarine That Wouldn't Come Up." *American Heritage*, 1958.
"Singer, E. C. Improvements in Sewing Machine." *Scientific American*, November 26, 1859.
Smyth, Augustine. "Torpedo and Submarine Attacks on the Federal Blockading Fleet off Charleston during the War between the Sections." *University of Virginia* (1951).
"South Carolina Confederate Twins." *Confederate Veteran* 33, no. 9 (September 1925).
Speer, William. "The Sub with Six Lives." *South Carolina Magazine*, February 1948.
Stanton, C. L. "Submarines and Torpedo Boats." *Confederate Veteran* 22, no. 4 (April 1914): 398–99.
Still, William N. "Confederate Naval Strategy." *Journal of Southern History* (August 1961).
"Submarine Engines." *Army and Navy Journal* (March 19, 1864).
"The Submarine Infernal Machine." *Scientific American*, August 17, 1861.
"Submarine Torpedoes—Infernal Machines." *Scientific American*, March 15, 1862.

Thatcher, Joseph. "The Courtenay Coal Torpedo." *Military Collector and Historian* 11 (Spring 1959).
Thomson, D. W. "Three Confederate Submarines." *U.S. Naval Institute Proceedings* (January 1941).
Thompson, Ray. "Where the Submarine Was Born." *Down South Magazine*, August 1958.
Tomb, J. H. "The Last Obstruction in Charleston Harbor, 1863." *Confederate Veteran* 32 (1924).
———. "Submarines and Torpedo Boats, C.S.N." *Confederate Veteran* 22, no. 4 (April 1914).
"Torpedoes Used by the Rebels." *Scientific American*, July 2, 1864.
"Treasury of Early Submarines (1775–1903)." *U.S. Naval Institute Proceedings* (May 1967).
Villard, O. S. "The Submarine and the Torpedo in the Blockade of the Confederacy." *Harpers Monthly Magazine*, June, 1916.
Von Kolnitz, H. "The Confederate Submarine." *U.S. Naval Institute Proceedings* (October 1937).
Von Scheliha, V. "Torpedoes in Warfare." *Army and Navy Journal* (May 8, 1869).
Wilkinson, D. "Peripatetic Coffin." *Oceans* 4 (1978).
"Wreck of the 'Otsego,' and the Explosion of the Tug 'Bazley' in the Roanoke River." *Harpers Weekly*, January 21, 1865.

Newspaper Articles

"Alabama's Contribution to Sub War." *Montgomery Advertiser*, May 9, 1959.
Alexander, William A. "The True Stories of the Confederate Submarine Boats." *New Orleans Picayune*, June 29, 1902.
"An Appeal to the Men of Texas." *Houston Tri-Weekly Telegraph*, May 26, 1865.
Arthur, S. C. "Early New Orleans Submarine Sired Davids." *New Orleans Picayune*, June 14, 1942.
"Builder of First Submarine Dead." *Mobile Register*, May 14, 1914.
"Captain McElroy Gives History of U-Boat Used in Confederate Navy." *Mobile Register*, November 5, 1924.
"Chaplain Builds Model of *Hunley*." *Mobile Register*, November 13, 1955.
"Coal Torpedoes." *New York Herald*, January 17, 1876.
"The Confederacy's Infernal Machine." *Mobile Register*, April 13, 1986.
"Confederate Sub Search Called Off." *Mobile Register*, June 27, 1981.
"Confederate Submarine Sunk a Warship but Herself Was Sunk Four Times." *Arkansas Gazette*, September 16, 1917.
"Confederates Built First Successful Submarine Boat." *Montgomery Advertiser*, May 26, 1907.
"Confederates Revolutionized Naval History." *Montgomery Advertiser*, September 14, 1950.
Craighead, Edwin. "The *Hunley*." *Mobile Register*, September 14, 1924.
"Davis and Texas." *Houston Tri-Weekly Telegraph*, May 24, 1865.
"Destruction of Steamer *Maria*." *Missouri Republican*, December 12, 1864.
"Did You Know the First Submarine Was Built Right Here in Mobile?" *Mobile Press Register*, April 11, 1948.
"Dixon, Builder of the Submarine *Hunley*, Went to Death in the Deep." *Mobile Daily Herald*, November 15, 1904.
Dunigan, T. "Descendants of Builder Unveil Plaque." *Mobile Press Register*, July 7, 1957.
East, Cammie. "Recalling H. L. Hunley." *Mobile Register*, October 23, 1988.
Erber, Henry. "Siege of Charleston, *Housatonic* Blown Up." *New York Times*, January 24, 1897.
"First American Ship Ever Sunk by a Submarine." *St. Louis Post Dispatch*, February 18, 1917.
"First Submarine Was Made in Alabama." *Montgomery Advisor Journal*, February 19, 1961.

"First Torpedo Boat." *New Orleans Picayune*, April 2, 1909.
"First War Submarine Built Here." *Mobile Register*, November 24, 1948.
"Forty Sailors Died in Iron Witch." *Mobile Register*, April 6, 1978.
Foster, J. "Is the Submarine in the Arcade of the Presbyter Really the *Pioneer*?" *New Orleans Picayune*, May 14, 1961.
"Fretwell Obituary." *Mobile Register*, April 19, 1885.
"George Dixon's Submarine, Details of How Brave Man Lost His Life." *Mobile Daily Item*, April 26, 1910.
"Group Trying to Preserve Birthplace of Sub *Hunley*." *Mobile Register*, September 22, 1962.
"The Gunboat *De Kalb* Blown Up by a Torpedo." *New York Times*, July 22, 1863.
"Gunboat *De Kalb* Blown Up by Torpedoes." *Philadelphia Enquirer*, July 22, 1863.
Hartwell, J. I. "An Alabama Hero." *Montgomery Advisor*, March 11, 1900.
Hearin, E. S. "*Hunley* Focal Point of Civil War Gathering." *Mobile Register*, February 2, 1989.
"Henry J. Leovy, Speaker of the Day." *New Orleans Times-Democrat*, May 13, 1900.
Hill, H. N. "Texan Gave World First Successful Submarine Torpedo." *San Antonio Express*, July 30, 1916.
"Historians Tell Exploits of Sub Built in Mobile." *Charleston Post and Courier*, January 29, 1943.
"The Hunley Building Dedicated." *Port of Mobile*, November 1964.
"Hunley's Niece Here." *Montgomery Advertiser*, September 6, 1956.
"*Hunley–Pioneer* Puzzles." *New Orleans Picayune*, December 4, 1967.
"*Hunley* Sub History Now at Library Here." *Mobile Register*, September 17, 1967.
"Importance of Capturing the Rebel Chiefs." *New York Times*, May 1, 1865.
Johnson, H. "Jeremiah Donivan, Survivor of *Hunley*, Died in His Bed." *Mobile Register*, November 19, 1948.
"Last Honors to a Devoted Patriot." *Charleston Mercury*, Monday, November 9, 1863.
"Launching of the *Hunley*." *Mobile Register*, November 6, 1914.
"Leovy, Henry J. Obituary." *New Orleans Picayune*, October 4, 1902.
"Letter to the Editor." *Mobile Advertiser*, November 11, 1924.
"Lieutenant Dixon." *Richmond Enquirer*, March 3, 1863.
Little, R. H. "The First Submarine to Sink a Hostile Warship." *Chicago Tribune*, November 29, 1936.
"Loss of the *Housatonic*." *New York Times*, February 27, 1864.
"Loss of the Sloop-of-War *Housatonic* by a Torpedo." *Chicago Tribune*, February 27, 1864.
Maury, R. "The First Marine Torpedoes Were Made in Richmond, V.A., and Used in James River." *Richmond Times Dispatch*, February 14, 1904.
McDonnell, Harry. "The Exploit of the Submarine *Hunley*." *Mobile Register*, August 3, 1969.
"Mobile-Built *Hunley* Wrote Chapter One in Submarine War a Century Ago." *Mobile Register*, February 16, 1864.
"Mobile Honors Crew of 1864 Submarine." *New York Times*, April 26, 1948.
"Mobile Honors Nine Pioneers of Submarine." *Birmingham News*, April 22, 1948.
"Monument Honoring *Hunley* Crew Unveiled at National Cemetery." *Mobile Register*, February 19, 1989.
Murphy, Alice. "Rebel Leader Discovered by Kin." *Houston Post*, July 13, 1967.
"Nautilus Gets Papers on Confederate *Hunley*." *Charleston News and Courier*, January 25, 1958.
"Navy Still Plans Hunt for *Hunley*." *Charleston News and Courier*, August 28, 1957.
"Other Submarine Boats." *Charleston News and Courier*, February 10, 1897.

Pinkerton, William. "Submarines of Civil War Met Disaster As Do Modern Craft." *Mobile Register*, June 11, 1939.
Price, Robert. "Submarine Looks Back on 75 Years." *Montgomery Advertiser*, February 17, 1939.
Rasco, B. "Two Cranes in Use Here Probably Lifted *Hunley*." *Mobile Press Register*, February 15, 1959.
"Rebel Submarine Machine Sunk." *New York Herald*, September 30, 1863.
"Records Disclose First Submarine Built at Mobile." *Mobile Register*, December 12, 1937.
"Remarkable Career of a Remarkable Craft." *Charleston Daily Republican*, October 8, 1870.
"Salvage of Submarine *Hunley* Has Chance for Success." *Charleston News and Courier*, June 18, 1957.
"Scott, John K. Obituary." *New Orleans Picayune*, March 10, 1874.
"Seven Men Were Drowned in Charleston Harbor by a Submarine Boat." *Gulf City Home Journal*, October 10, 1863.
Smith, F. G. "Submarine Warfare." *Mobile Register*, June 26, 1861.
Spotswood, James. "Builders of First Submarine to Be Honored in Mobile." *Montgomery Advertiser*, April 21, 1948.
Stuart, Ben. "First Submarine Torpedo Vessel." *Galveston Daily News*, August 29, 1909.
"The Submarine Boat Which Is Thought to Have Been the First of Its Kind." *Mobile Register*, February 3, 1895.
"Texas and the War—the Untrodden Domain of the South." *New York Times*, June 1, 1865.
"Today to Recall First Ship Sunk by Sub's Torpedo." *Mobile Register*, January 3, 1951.
"Veteran of Three Wars Answers Last Call." *Mobile Register*, November 19, 1928.
Wainwright, T. "First Submarine to Send Man of War to Davy Jones' Locker Was Built in Mobile." *Mobile Register*, August 30, 1931.
"Yankee Gun Boat Found." *Mobile Register*, July 19, 1981.
Zeigler, Mary "S.C. Leaders to Attend Dedication of *Hunley*." *Charleston News and Courier*, July 12, 1967.

Index

Abbeville, South Carolina, 172–73, 176, 178
Alexander, William: background and overviews, 26; breech-loading cannon design, 88–89, 140; on dangers to *Hunley* crew, 85; on description and plans for *Hunley,* 28–30; Freemasonry, 192; on *Hunley* crew recruitment, 59, 75–76; post-war life, 188–89; on practice attacks, 88; on sinking of *American Diver,* 27; on sinking of *Housatonic* and *Hunley,* 103; on sinking of *Hunley,* Oct., 1986, 68–69; on spar torpedo practice, 84; on submerged endurance test, 86–87; on torpedo problems and strategies, 76–77; transfer to Mobile order, 88–89. *See also in* gallery, #23
Alexandria, Louisiana, 23
American Diver (submarine), 27, 130, 142–43
anchoring and deployment techniques, 19, 35, 159. *See also* mining strategies
Appomattox and Lee's surrender, 169
Army of Northern Virginia, 127, 147, 165
Atchafalaya Bay/River, 36
Atlanta, Georgia, siege of, 138–39, 157–58

Baker, J. T., 72
ballast strategies, 29
"Bands of Destruction," 33–34
Baron de Kalb, USS, sinking of, 41–42, 74, 114. *See also in* gallery, #9
Barrett, Theodore H., 180
Battery Marshall, 81, 84, 101. *See also in* gallery, #30
Battery Wagner, 47
Battle House Hotel, Mobile, 44, 57, 140
Battle of Monocacy, 127
Battle of Palmito Ranch, 180
Battle of Wilderness, 121
Bazely (Bezely), (Union tug), sinking of, 148. *See also in* gallery, #48

Beauregard, P. G. T.: on crew recruitment, 75–76; dangers of *Hunley,* response to, 72–73; final orders for *Hunley,* 85; *Hunley* move to Charleston, 45–47; retreat with Davis, 169; salvage of *Hunley* ordered, 57; seizure of *Hunley* for military crew, 52–53; on Sherman's movements, 157; on sinking of *Hunley,* February, 1864, 101; on sinking of *Hunley,* Oct., 1863, 62. *See also in* gallery, #15
Becker, Arnold, 76
Bee, Hamilton, 70
Belton, E. C., 85
Benjamin, Judah P.: and blame for Lincoln assassination, 171; bounty settlements, 136–37; efforts in retreat, 169; and fall of Richmond, 5, 166, 167; foreign agents, plan to contact, 173, 176–77, 178; relationship with Singer's Corps, 10
Blakely River, 162–63
blockade running, 26, 152–53
blue light signal, 92, 93–94, 98, 100
"Boat Burners," 136
Boggs, W. R., 106–7
Booth, John Wilkes, 171, 172
bounties/rewards: determination of, 136; Dillingham gold, 139, 140; evidence of payments, 185, 191; Long Bridge plot, 137–38; settling claims, 136–37; ships, 31, 42, 49, 51; steamboats, 134–35
Bradbury, David: background and overviews, 20; mining strategies, 3, 34–35, 71–72, 81–82, 115; on need for exclusive use of Close and Sons, 123–24; role in Singer's Corps, Port Lavaca, 21, 22, 70; torpedo boat construction, Houston, 121, 144
Bradford, R. B., 128
Bradfute, W. R., 71
Bragg, Braxton, 65, 158, 176

Braman, John D.: background and overviews, 20; Charleston, move to, 48; cigar boat design, 58, 67, 74–75; documents of, lost and captured, 107–9, 111, 113; Mobile, activities in, 20, 32; orders back to Texas, 125; postwar life, 186; torpedo boat construction, Houston, 152, 175–76; on torpedo boat construction, 182; on York river torpedo strikes, 78
Brazos Santiago, Texas, 70
Breach Inlet mooring, 81, 83, 93–94. *See also in* gallery, #30
Breckinridge, John, 167, 176
Broadfoot, David, 57, 63
Brown, Isaac, 31, 40–41, 42, 45
Brownsville, Texas, 70, 180
Bryan, E. Pliny, 125–26
Buchanan, Franklin, 44–46, 130
Buffalo Bayou torpedo boat construction, 18–19, 121, 122, 150–51, 175. *See also under* Houston, Texas
"Bummers, Sherman's," 158
Bumpass, Alexander J. "Tip," 219*n* 1
Bushnell, David, 118–19
Butler, Benjamin F., 11, 12, 118, 119

Cairo, USS, sinking of, 23, 31, 34
Canandaigua, USS, 97–98, 99, 101
Canby, E. R. S., 181
Cane, Michael, 55
Cape Fear River, 154, 158–59
Carlson, C. F., 91
Charleston, South Carolina: blockade of, 152, 154; defensive strategies, 47; *Hunley*, naval seizure of, 52–55; *Hunley*, sinking of, 43, 62–63; *Hunley* move to, 45–48; *Hunley* operations in, 48–52; relaunch of *Hunley*, 76. *See also Housatonic*, USS, sinking of; *in* gallery, #16, 17
Charleston Daily Courier, 51–52, 55, 92
Charleston Daily Republican, 83
Charleston Mercury, 69, 102–3. *See also in* gallery, #32
Charlotte, North Carolina, 171–72
Chenango (Union steamship), 136
"Chicago Remnant," 137
Chicora, CSS, 50, 54–55

Chubb, Thomas H., 150–51, 156–57
Chubb's shipyard, Galveston, 156–57, 161, 174, 182
cigar boats (*David*-class design), 67, 74–75, 108, 114. *See also in* gallery, #21, 28; torpedo boats (*David*-class design) overviews
Clark, Henry E., 107, 110–11
Clarke, Thomas L., 136–37, 139–40
Clingman, Thomas, 52–53
Close and Sons fabricators, 123–24, 144, 150–51
coal torpedoes, 5, 110–11, 112, 136, 139, 141
Collins, F., 76
Confederate Engineering Office/Department, 61–62, 77, 111–12
Confederate Secret Service, 5, 24, 34, 61, 110. *See also* Courtenay's Secret Service Corps; McDaniel's Secret Service Corps; Sage, Bernard Janis; Singer Secret Service Corps
Congdon, Joseph, 97
Cook, Joseph, 14
copper torpedo casings, 175
Corpus Christi, Texas, 70–71
cost and funding of *Hunley*, 28, 30
"cottonclads," 17
Courtenay, Thomas E., 5, 110–11, 136
Courtenay's Secret Service Corps, 34, 134, 135, 139
Court of Inquiry accounts: conclusions of, 104–5; Congdon, 97–98; Craven, 96–97; Crosby, 95, 98; Fleming, 98; Green, 97–98; Higginson, 95, 96; Jones and Hassenger, 217*n* 62; Murphy, 140–41, 216*n* 26, 216*n* 34; Pickering, 93, 95–96
covert operations. *See* sabotage groups
crank, hand. *See* manual propulsion
Craven, Charles, 96–97
Craven, Tunis, 130–31
Crosby, John, 94, 95, 98, 99
Cussler, Clive, 192. *See also in* gallery, #51

Dahlgren, John, 80, 82–83, 85, 99–100, 100–101. *See also in* gallery, #22
"Damn the torpedoes..." (Farragut), 131
Dantzler, O. M., 94, 98, 100
Danville, Virginia, 167

David (torpedo boat), 58, 61, 77, 80, 82–83. *See also* cigar boats (*David*-class design)
Davis, George, 167, 172
Davis, Jefferson: capture of, 178; efforts in retreat, 167, 169–73; and electro-magnetic engine request, 143–44; and fall of Richmond, 5, 165–67; Long Bridge plot, 137–39
Davis, Nicholas, 55
deadlights, 160
Dekalb (ship). *See Baron de Kalb,* USS, sinking of
Demonstrations: *Hunley,* Mobile, 44–45, 46; *Hunley,* Richmond, 43; torpedoes, 16, 18–19, 43, 61–62
deployment techniques, 19, 35, 159. *See also in* gallery, #26
depth monitoring, 29, 44
deserters (Confederate), 79–80, 85, 152, 156, 161–62
detonation strategies/mechanics, 43, 84, 127–29, 160. *See also in* gallery, #34
Dillingham, Henry: background and overviews, 4–5; and blame for Lincoln assassination, 171, 187–88; covert operations, 135, 136–37; gold ingot bounty, 139, 140; post-war life, 187–88; third crew of *Hunley,* 59; transfer to Mobile order, 89
dimensions of torpedo boats, 28–30, 54, 160
"Diver," 83, 85. *See also H. L. Hunley (Hunley)*
divers, salvage, 57, 63
Dixon, George E.: background and overviews, 4, 30, 59–60, 73; crew recruitment, 72; final attack and loss of *Hunley,* 94–98; post-war estate, 189, 191–92; return of *Hunley* to Singer's group, 59; will, 73. *See also in* gallery, #37
documents (Confederate): cache of in retreat, 176, 178; captured, 33, 90–91, 107–9, 110–13, 115; destruction of, 23, 166, 176
Dog River, 132
Donovan, Jeremiah, 48
Doyle, Frank, 55
drift mines/torpedoes, 83, 119–20
Dunn, Robert W.: background and overviews, 20; Charleston, move to, 48; cigar boat design, 58, 67, 74–75; documents of, lost and captured, 110–13; introduction letter to Magruder, 114–15; mining strategies, Port Lavaca, 157; Mississippi crossing, 107, 109–10; post-war life, 189; Sabine River mining, 123; steam engine procurement, 149; on team's accomplishments, 36; torpedo boat construction, Trans-Mississippi theater, 121, 122–23, 144, 174–75; Wilmington request for torpedo boats, 77–78; on York river torpedo strikes, 78–79

Eagle, Henry, 12, 13
Eastport, USS sinking of, 115–16
Echols, John, 171–72
electric motors, 4, 26–27, 142–44, 153–54
Engineer Troops, 24, 32–33
Erwin, William R., 138, 140, 141
Etiwan/Etewan (steamer), 54–55
Ewing, Francis, 34

Farragut, David G., 9–11, 12, 13, 129, 130–31, 133
Fayetteville, North Carolina, 157–59
Fayetteville Weekly Intelligencer, 104. *See also in* gallery, #37
Fleming, Robert, 98
Fletcher's house, 140
floating mines/torpedoes, 83, 119–20
Florida, Leovy's retreat to, 178
Ford, John Salmon "Rip," 180
foreign agents, plan to contact, 173, 176–77, 178
Fort Branch, 149
Fort Donelson, 9
Fort Esperanza, 12–13, 21, 34–35, 71–72, 81–82
Fort Fisher, 154
Fort Henry, 9
Fort Hugar, 133
Fort Jackson, USS, 181
Fort Johnson, 50, 54
Fort Morgan, 26, 130, 132
Fort Moultrie, 47
Fort Stedman, 165
Fort Sumter, 47, 52, 53. *See also in* gallery, #31
Frary, C. E., 20–21, 48, 186
Frazier, Edward, 136–37, 137–39, 139–40, 171, 187–88
Freemasons, 19–20, 185, 191–92. *See also in* gallery, #50

Fretwell, John R.: background and overviews, 2, 16; defense of Mississippi waterways, 31; deployments to North Carolina, 146–47, 157–59; Mobile, arrival in, 27–28; post-war life, 185–86; Richmond, activities in, 118, 119–20, 125–27

Fretwell's Percussion Torpedo, 128

Galveston, Texas: capture of, 12, 13–14; mining strategies, 22; profile, mid 1800s, 8; recapture of, Confederate, 17–18; submarine/torpedo boat construction, 3, 144, 156–57, 161, 174, 182; torpedo factory, 2

Galveston Daily News, 182

Gertrude, USS, 152

Gilmer, J. F., 24, 132, 144, 146

gold coin legend, 73

Grant, Ulysses S.: Lee's surrender to, 169; on railroad sabotage, 67; Richmond offensives, 118, 120–21, 165–67; taking of Vicksburg, 39–40; in Tennessee, 64; Yazoo City, advance on, 40–42

Gray, M. M., 56

Gray Cloud (Union transport), 36, 114

Green, Joseph F., 97–98, 99, 101

Greensboro, North Carolina, 169

Greyhound, USS, destruction of, 5

Guadalupe River, 72, 82

Halligan, John P., 34

Hartford, USS, 131

Hasker, Charles H., 54–55

Hassenger, C. M., 152, 217*n* 62

Haviland, James E., 121

Hébert, Paul, 9, 17

Heroes of America, 142, 156

Herron, F. J., 40, 41

Higginson, F. J., 94, 95, 96, 99–100

Hill, H. N., 25

Hill, Horace, 189–90

Hirshberger, L. C., 21

H. L. Hunley (*Hunley*): attack, final, 94–98; attack plan, final, 92–94; Charleston operations and exercises, 45–48, 76–77, 80–81, 84–89; construction in Mobile, 24; construction of, 25, 26, 28, 59, 140; cost and funding, 26, 28, 30; demonstration, Mobile, 44–45, 46; demonstration, Richmond, 43; description and operation, 28–30, 38, 44, 54, 56, 79; endurance test, 86–87; excavation and recovery, 6; launch, 44; model of, by deserters, 85; naval seizure of (Confederate), 52–53; recovery/repair of, Aug.,1863, 45–48; recovery/repair of, Oct.,1863, 63–64, 68; recovery/repair of, Oct.–Nov.,1863, 74, 79; research and discoveries on, 6; sinking of, Aug., 1863, 54–55; sinking of, February, 1864, 98, 99–105; sinking of, Oct., 1863, 62–63, 68–69, 79. *See also in* gallery, #12, 13, 27, 29, 38, 39, 40

Hood, John Bell, 151

"Horse Marines," 17–18

"Horsemen of the Sea," 17–18, 168

Housatonic, USS, sinking of, 83, 92, 94–98, 99–104, 114. *See also in* gallery, #35, 36

Houston, Sam, 8

Houston, Texas: Galveston retreat to, 13–14; submarine/torpedo boat construction, 3, 18–19, 121, 144, 149–51, 175–76; torpedo demonstration, 18–19

Hunley, Horace Lawson: burial and funeral, 69–70; Charleston, move to, 48; death in sinking of *Hunley,* 62–63, 68–69; defense of Mississippi waterways, 31–32, 39; funding of submarine construction, 26; Mobile, activities in, 25, 27–28, 57; naming of *Hunley,* 5; *Pioneer,* 10; request for command of *Hunley,* 58–59; will, 39. *See also in* gallery, #6, 24, 32

Hunnicutt, M. P., 151, 154–56, 159–61

Hunt, Edward B., 220*n* 44

Hutchinson/Hutcheson, Ike R., 122, 159

Indian Chief, CSS, 62, 75–76, 79, 85

Indianola, Texas, 13

ironclads (ships), 51, 52–53. *See also in* gallery, #19

Ironsides, USS, 49, 51, 54–55, 83. *See also in* gallery, #18

Jackson, A. M., 160

Jackson, Thomas J. "Stonewall," 17

James River: mining strategies, 43–44, 118–20, 125–27; torching of railroad bridge, 167; torpedo demonstrations, 43, 61–62

Johnson, Andrew, 171
Johnston, Joseph E. (Gen): Mississippi command, 35, 40, 42, 114; North Carolina command, 165, 169–70
Johnston, Sidney, 9
Jones, Henry, 152, 217*n* 62
Jones, James: background and overviews, 20; and plot to destroy *Tennessee*, 155; post-war life, 186; Red River operations, 159; return to Shea's battalion, 145; Richmond torpedo operations, 43; torpedo demonstration, 61–62; Wilmington torpedo boats, 78
Jordan, Thomas, 57, 59, 74, 88

keg torpedoes, 56, 125–26
Kelly, John, 55

land mines, 50, 66, 71–72, 120, 146. *See also* railroad sabotage
last battle of Civil War, 180
Lavaca Guards, 16
Leadbetter, Danville, 38
Lee, Francis, 83–84
Lee, Robert E., 2, 43, 120, 126, 165, 169
Lee, S. P., 118
Leovy, Henry: as colonel of cavalry, 156; as commissioner for southwestern Virginia, 28, 142; and confirmation of *Hunley* sinking (1864), 105; escape from Richmond, 5; foreign agents, plan to contact, 173, 176–77, 178; Mobile, activities in, 25, 27–28; *Pioneer*, 10; post-war life, 187; retreat to Charlotte, 171–73; surrender of, 178
lighting strategies, 160, 161
Lincoln, Abraham: assassination, 5, 171, 187–88, 190; Grant, support of, 121; second term election, 147, 213*n* 95
liquor, destruction of military stores, 166–67
Long Bridge plot, 137–39
longevity of underwater torpedoes, 35
Longnecker, William, 20, 22
Louisiana: covert operations in, 106; mining of waters in, 36, 115–16, 123; New Orleans, capture of, 10–11, 129–30; Red River, 121, 122, 125, 159–61; Shreveport, 115–16, 121, 125, 159, 160, 161; Smith's headquarters in, 23. *See also* Mississippi River

Lubbock, Francis, 14
Lubbock, Henry S., 124, 175–76
Lubbock's Mill, 150, 175, 181
Lynchburg, Texas, 157, 217–18*n* 64

Macomb, W. H., 147–49
Magruder, John B. "Prince John": coastal refortifications, 42; command at Houston, 17; late war efforts, 174–75; and retaking of Galveston, 17–18; surrender of, 181; and Union invasion of Texas 1863, 70–71
Mallory, Stephen R., 141, 167, 168, 169, 176
Manassas, Virginia, 9
manual propulsion, 27, 29, 77, 160
marine worms, 127–28
Marshall, Texas, 174
Masonic Lodge members, 19–20
Matagorda Bay/Island, 9, 12–13, 14. *See also* Port Lavaca, Texas
Maury, Dabney, 131, 162. *See also in* gallery, #7
Maury, Matthew, 24, 44, 188
McClellan, George, 213*n* 95
McClintock, James: on crank propulsion, 27; electric powered submarine, 4, 26–27, 142–43; *Hunley* command, 52–53; Mobile, activities in, 25–26, 27–28, 127–28, 129–30, 132, 163; *Pioneer*, 10–11; post-war life, 188. *See also in* gallery, #6
McDaniel, Zedekiah, 34, 120
McDaniel's Secret Service Corps, 34, 117–18, 134, 135, 139
McGill, Thomas, 142
Mexico: Confederates flight to, 181, 183; Hunnicutt in, 159; wartime trade, 169, 170
Middleton, Harriet, on *Hunley*, 49, 51, 53–54, 55, 103
military credentials for Singer's Corps, 32, 33–34, 50, 66, 117–18, 123. *See also in* gallery, #24
Milwaukee, USS, sinking of, 163
mines. *See* coal torpedoes; land mines
mining strategies: Charleston, 47; Fayetteville, 159; Fort Esperanza, 3, 21, 71–72, 81–82; Galveston, 22; Guadalupe River, 72, 82; James River, 43–44, 118–20, 125–27; Louisiana waters, 36, 123; Mississippi River, 142; Mobile area, 24, 130–31, 132, 162; Port

mining strategies (*continued*)
  Lavaca, 21–22, 34–35, 81–82, 157; Red River, 159; Roanoke River, 147–49; Sabine River, 123; Shreveport, 115–16; Tallahatchie River, 31; Yazoo River, 31, 40–42. *See also* railroad sabotage
Mississippi (State), 31, 39–40, 40–42. *See also* Mississippi River; Trans-Mississippi Department; Yazoo River/shipyard
Mississippi River: mining strategies, 142; New Orleans, capture of, 10–11, 129–30; sabotage activities, 5, 134–35; Union capture of, 2–3, 40–43. *See also* Louisiana; Mississippi (State); Trans-Mississippi Department
Mobile, Alabama: abandonment of, 162–64; torpedo demonstrations, 44–45, 46; torpedo design improvements, 127–28. *See also* Mobile Bay
Mobile Advertier, 103
Mobile Bay: mining strategies, 24, 32, 38, 130–31, 132; surrender of, 129–32. *See also* Fort Morgan; *in* gallery, #8
Mobile Daily Herald, 73
Mobile Register, 104
Morris Island, 47, 51–52. *See also in* gallery, #20, 31
Mount Pleasant mooring, 77, 79, 80. *See also in* gallery, #31
Munsey's Magazine, 75, 88, 89
Murphy, William, 140–41, 216*nn* 26, 216*nn* 34

National Underwater and Marine Agency (NUMA), 225*n* 41
Native American Confederates, 168, 181
New Orleans, Louisiana, capture of, 10–11, 129–30
New Orleans Picayune, 88, 89
New York Times, 169
Niblett's Bluff, Louisiana, 123
Noland, Bill, 135

Osage, USS, sinking of, 163
Otsego, USS, sinking of, 147–49
ownership of *Hunley,* 20, 73–74, 78, 105, 109, 196*n* 90
Ozark, USS, 159

Palmetto State, CSS, 50–51
Palmito Ranch, Battle of, 180
Park and Lyons machine shop, Mobile, 25, 26, 28, 59, 140. *See also in* gallery, #10
Parks, Thomas, 59, 62, 68–69. *See also* Park and Lyons machine shop, Mobile
Pass Cavallo, 12, 21, 35, 71
patents, torpedo, 35, 49, 58
payments/wages: for contracted explosive devices, 22, 31, 32, 35, 38, 44, 109; for covert operations, 34, 135, 137; for torpedo boat construction, 150–51; wages for Singer's Submarine Corps, 25. *See also* bounties/rewards; *in* gallery, #11, 14
Payne, John A., 53, 54–55, 57–58
Pearce, Nathanial, 219*n* 1
percussion torpedoes, overviews, 19, 128. *See also in* gallery, #4
"Peripatetic Coffin," 53
Petersburg, Virginia, 147
Phillips, J. L., 168, 174, 179
physical demands on *Hunley* crew, 86
Pickering, Charles, 92, 93, 94–95, 95–96, 99
Pinkerton, Allen, 61
Pioneer (submarine), 10–11, 25, 130. *See also in* gallery, #5
Polk, Bishop, 135
Pope, John, 179
Porter, David D.: blockade of Shreveport, 125; on captured documents, 111–12, 141; on coal torpedoes, 141; Red River Campaign, 115; Yazoo City, capture of, 40–42
Port Hudson, Louisiana, 36, 40
Port Lavaca, Texas: defensive operations, surface, 12–13; founding of Singer Secret Service Corps, 1; Matagorda Bay/Island, 9, 12–13, 14; mining strategies, 21–22, 34–35, 81–82, 157; Pass Cavallo, 12, 21, 35, 71; siege of, 14–15; Singer's move to, 9; torpedo construction, contract, 19, 21. *See also* Fort Esperanza; *in* gallery, #2; *in* gallery, #3
propulsion methods: electric motors, 4, 26–27, 142–44, 153–54; manual, 27, 29, 77, 160; steam engines, 27, 149

"Quaker guns," 13

Railroad Bridge (Weldon) plot, 2, 147–49
railroad sabotage, 50, 64–67, 114
Rainbow Bluff, 147–49
Rains, Gabriel: Charleston defense, 47; on Fretwell's defense of Fayetteville, 159; on Fretwell torpedo, 128–29; mining strategies, 118–19; Richmond defense, 125–27; on sabotage successes, 146, 185; Torpedo Bureau, oversight of, 56
ranks and military credentials, 32, 33–34
Rattlesnake Shoal, 91, 92
Reagan, John, 167
records. *See* documents
Red River, 121, 122, 125, 159–61. *See also* Shreveport, Louisiana
"The Red River Campaign," 107, 113, 115, 127
Renshaw, William, 13
"The Republic of Republics" (Sage), 187
research and methodology, 6–7
Revolutionary War, 118–19
Rhodes, Rufus, 35, 49, 58
Richmond, USS, 36
Richmond, Virginia, 43, 61–62, 119–20, 125–27, 165–67. *See also* James River
Roanoke River, 147–49. *See also in* gallery, #48
rocket-powered torpedoes. *See* self-propelled torpedoes
Rock Fish Creek, 159
Roddey, P. D., 65

Sabine River, 123
sabotage groups: and captured Confederate documents, 112–13; late war activities, 134–42; locations/postings of operatives, 33, 106, 139, 177; post-war myth and secrecy, 190; Sage's support of, 24–25, 33, 106–7, 134. *See also* Confederate Secret Service; railroad sabotage
Sage, Bernard Janis, 24–25, 33, 106–7, 134, 187
salt water issues, 35, 127–28, 131–32
San Antonio Express, 18, 21
Savannah, Georgia, 35–36
Scientific American, 143
Seddon, James, 9, 24, 136, 144
self-propelled torpedoes, 3, 67, 122, 220–21n 44
Selfridge, Thomas O., 115–16
Selma, Alabama, 3, 41, 42

semi-submersibles, 58–59, 80
sewing machine, 9
Shea, Daniel D., 9, 15, 22
Shea's Battery of Texas Light Artillery, 1, 9, 12–13, 14–15
Sheridan, Phillip, 182–83
Sherman, William T., 66, 138–39, 157–59, 169–70
Shreveport, Louisiana, 115–16, 121, 125, 159, 160, 161
signal system, 92, 93–94, 98, 100
Simkins, C., 76
Singer, Edgar Collins: background and overviews, 1; Charleston, activities in, 48; family and move to Texas, 8–9; letter to wife, (captured), 90–91; Mobile, arrival in, 27–28; orders back to Texas, 125; personal appearance, 4, 9; post-war life, 189–90; on torpedo boat construction, Trans-Miss theater, 142; Wilmington request for torpedo boats, 77–78. *See also in* gallery, #1
Singer, Isaac Merritt, 8–9
Singer Secret Service Corps (Singer's Corps): activities of, synopsis by Dunn, 114–15; founding/origin, 1, 34; locations/postings of operatives, 106, 139, 177. *See also in* gallery, #45; parole of, 183 (*See also in* gallery, #49, 50); post-war lives, 185–89; recruiting early members, 19–21; research and discoveries on, 6; sabotage activities, 64–67, 134–42; transfer of members to War Department, 117–18; at war's end, 181
Singer('s) Submarine Corps, 25. *See also* Singer Secret Service Corps
Singer's Torpedo Company. *See* Singer Secret Service Corps
sinking of Union vessels: *Baron de Kalb,* 41–42, 74, 114; *Bazely (Bezely),* 148 (*See also in* gallery, #48); *Cairo,* 23, 31, 34; *Chenango,* 136; *Eastport,* 115–16; *Gray Cloud,* 36, 114; *Greyhound,* 5; *Housatonic,* 83, 92, 94–98, 99–104, 114 (*See also in* gallery, #35, 36); *Milwaukee,* 163; Mobile area, 163; number of sinkings, 185, 191; *Osage,* 163; *Otsego,* 147–49; Roanoke River, 147–49 (*See also in* gallery, #48); *Sultana* (Union transport), 5; *Tecumseh,* 1, 130–31, 132 (*See also in* gallery, #46)

Sleeman, C. W., 149
Smith, Angus, 57, 63–64
Smith, E. Kirby: background and overviews, 3; escape to Mexico, 183; and Sage's covert operations, 106; Singer's Corps reporting to, 23; surrender negotiations, 179–80, 181; torpedo boat construction order, 121. *See also* Trans-Mississippi Department
Spanish Fort, 133, 162–63. *See also in* gallery, #8
spar-mounted torpedoes, 67, 83–85. *See also in* gallery, #27, 33
speed of torpedo boats, 160
Spotswood Hotel, Richmond, Virginia, 2, 61. *See also in* gallery, #43
Sprague, Charles L., 53
The Spy of the Rebellion (Pinkerton), 61
Stanton, C. L., 54, 62
steamboats, sabotage of, 134–35, 136
steam engines, 27, 149
steering strategies, 29, 30
"The Subterranean Room," 61
subterra shells, 120, 146. *See also* land mines
Sullivan's Island mooring, 80–81
Sultana (Union transport), sinking of, 5
supply trains (Union), sabotage of, 50, 64–67
"The Swamp Angel," 51. *See also in* gallery, #25

Tallahatachie River, 31
Taylor, Richard, 106, 165, 176–77
Tecumseh, USS, sinking of, 1, 130–31, 132. *See also in* gallery, #46
Tennessee, CSS, 130, 132, 133
Tennessee, sabotage in, 64–67. *See also* Long Bridge plot
Tennessee, USS, 155, 160
Texas: Davis's planned retreat to, 170–71, 171–72, 174; effect of war on, 169; Lynchburg, 157, 217–18*n* 64; Shea's Battery of Texas Light Artillery, 1, 9, 12–13, 14–15; Union invasion of, 1863, 70–71; Union invasion of, 1865, 180. *See also* Galveston, Texas; Houston, Texas; Port Lavaca, Texas
"Texas Gave World First Successful Submarine Torpedo" (Hill), 189–90
Texas Ordinance of Secession, 8
Texas Secession Convention, 8
Tidwell, William, 34

time, submerged, 56, 86–87
Tomb, James, 77, 80, 83, 98
torpedo boat construction: Galveston, Texas, 3, 144, 156–57, 161, 174, 182; Houston, Texas, 3, 18–19, 121, 144, 149–51, 175–76; Trans-Mississippi theater overviews, 116–17, 121, 122–23, 144, 174–75
torpedo boats (*David*-class design) overviews, 58, 77–78, 80–81, 89–90, 100–101. *See also* cigar boats (*David*-class design)
Torpedo Bureau, 22–23, 38, 56, 127
Torpedoes: coal torpedoes, 5, 110–11, 112, 136, 139, 141; drifting, 83, 119–20; Fretwell design, 1864, 128–29, 148; keg, 47, 56; recovered/restored, 185; self-propelled, 3, 67, 122, 220–21*n* 44; spar-mounted, 67, 83–85 (*See also in* gallery, #27, 33); spy reports, 160; stationary, 19, 43, 71–72 (*See also in* gallery, #4, 42); towed, 44
trains, sabotage of, 50, 64–67, 114
Trans-Mississippi Department, 23, 168, 170–71, 173–74, 179–80, 181. *See also* Smith, E. Kirby
Treasury (Confederate), evacuation of, 166
Trenholm, George, 49, 55, 137, 167, 172
triggering mechanisms, 19
tripwire mines, 50
Tucker, F. M., 21, 45–46, 50–51, 80
Tucker, Joseph W., 135

underwater contact mines, overviews, 1–2. *See also in* gallery, #41, 47; torpedoes
underwater explosives, early experiments, 11, 15–16, 18–19
uniforms and military status, 50. *See also* military credentials for Singer's Corps
U.S. Naval Institute Proceedings, 8

Vance, Zebulon, 170
Van Dorn, Earl, 9
ventilation strategies, 29, 86
Vicksburg, Mississippi, 39–40
viewports, 44
Von Scheliha, V., 132

Walker, John G., 41, 144, 150–51
Walker, W. S., 81

Ward, (unk) Colonel, 107, 110
Warren Lasch Conservation Center, 6
Watie, Stand, 168, 181
Watson, Baxter: Charleston, move to, 45–47; electromagnetic engine, quest for, 4, 26–27, 142–44, 153–54; Mobile, activities in, 25–26, 27–28, 127–28, 129–30, 163; *Pioneer*, 10; post-war life, 186–87
Weldon Railroad Bridge plot, 2, 147–49
Welles (Wells), Gideon, 41, 85, 185
Western Gulf Blockading Squadron, 9–10
West Point Museum, 185
Whipple, W. D., 66–67
Whiting, W. H., 77–78
Whitney, B. A. "Gus," 20, 21, 32, 45–47, 53, 185
Wicks, James A., 76
Wigfall, Louis R., 74, 75, 77–78

Willey, Henry, 73
Williams, Absolum, 200*n* 66
Williams, James J., 47
Williams, John, 142
Wilmington, North Carolina: fall of Fort Fisher, 154; ironclad torpedo boat construction, 3, 77–78; torpedo contract, 35; Union plan to occupy, 152; Weldon Railroad bridge plot, 147–49
Wilson's Creek, Missouri, 9
Winnebago, USS, 130
Wood, Taylor, 178

Yazoo River/shipyard, 23, 31–32, 34, 39, 40–42, 114. *See also Baron de Kalb*, USS, sinking of
yellow fever, 13, 15